The Best American Science and Nature Writing 2005

GUEST EDITORS OF
THE BEST AMERICAN SCIENCE
AND NATURE WRITING

The Best American Science and Nature Writing 2005

Edited and with an Introduction
by Jonathan Weiner

Tim Folger, Series Editor

HOUGHTON MIFFLIN COMPANY
BOSTON · NEW YORK 2005

ISSN: 1530-1508
ISBN-13: 978-0-618-27341-6   ISBN-10: 0-618-27341-7
ISBN-13: 978-0-618-27343-0 (pbk.)   ISBN-10: 0-618-27343-3 (pbk.)

Printed in the United States of America

MP 10 9 8 7 6 5 4 3 2 1

Contents

Contents

Foreword

ONE FEBRUARY MORNING two years ago a loud noise rattled my house and woke me up. At first I thought a tree branch had fallen on the roof. Not until a few hours later did I realize that what I had heard was a death knell; it was a sonic boom from the doomed space shuttle *Columbia* hurtling over northern New Mexico, where I live, at twenty times the speed of sound in the final stage of its descent from orbit. Minutes later the shuttle disintegrated 205,000 feet above central Texas; all seven astronauts on board died. It was *Columbia*'s twenty-eighth mission.

Columbia made its maiden flight twenty-four years ago, on April 12, 1981. It was the very first shuttle to be launched, and the world's first winged spacecraft. (Ian Parker writes about the second such vehicle in "The X Prize," on page 202.) At the time, NASA officials promised that the new shuttle fleet would make space flight routine, with weekly launches. Just how wrong they were is made tragically clear in "Stumbling into Space," Timothy Ferris's contribution to this anthology.

The title of Ferris's article gives me pause. Stumbling into space? This isn't what I imagined the twenty-first century would be like when I was a child, rising early to watch the launch of *Mercury* and *Gemini* missions forty years ago. The moon landing was only a few years away. Arthur C. Clarke's vision of a gracefully wheeling, tourist-friendly space station and missions to Jupiter in *2001: A Space Odyssey,* seemed not just plausible but inevitable. President Kennedy called space "this new sea." But the anticipated space age never really arrived. The launching pad that sent the first men to the moon is rusting and surrounded by weeds. (A petition to save the site is available on the Web at http://www.petitiononline.com/LUT/petition.html.)

The challenge of space exploration, though, was and is self-imposed, a goal for an overachieving species. It's not something we need to do, at least not right away. But the century stretching before us holds challenges that no one dreamed of forty years ago, challenges that we have no choice but to confront. Unfortunately, many of those in a position to initiate effective policies on issues ranging from the environment to the treatment of disease have a shocking disdain for science. Bill McKibben, in "Crossing the Red Line," writes about the depressing failure of our nation's leaders to even acknowledge the looming threat of global warming. And in "Please Stand By While the Age of Miracles Is Briefly Suspended," James McManus describes how ill-conceived policies on stem cell research may doom his daughter, and millions of others with diabetes, to an early death from heart disease or kidney failure.

One of the strangest and most disturbing developments in the twenty-first century has been the surge in religious fundamentalism, both here and abroad. Who would have imagined that our society would still be rehashing the same issues raised by the Scopes trial in 1925? It should be a cause for national embarrassment, and Natalie Angier, in "My God Problem — and Theirs," with her characteristic pungent humor, takes the scientific community to task for not mounting a spirited and public defense of rationalism.

If it were within our power, no doubt we would have conjured a better beginning for the new millennium. Even so, the past year held many surprises that I wouldn't want to have missed, events that wouldn't have been out of place in the brightest of imagined futures. Within the past few weeks alone I was able to download and listen to the sounds made by the *Huygens* probe as it tore through the atmosphere of one of Saturn's moons; I watched a NASA video of dust devils swirling across a Martian desert when I should have been working on this foreword; I read how scientists had extracted soft tissue — including blood vessels — from the fossilized leg bone of a seventy-million-year-old *Tyrannosaurus rex;* and I've spent an enjoyable year helping to collect articles for this remarkably diverse anthology, of which I've mentioned only a few. The other stories range from Jared Diamond's essay on the collapse of civilizations to Oliver Sacks's incredible account of the mechanisms of consciousness. Jonathan Weiner, our guest editor, has more to say in his introduction about the selections he has made.

As I write, 2005 is well under way, and I'm already searching for articles for next year's anthology. Typically I forward nearly one hundred of the best articles I've read to the guest editor of the series, who then makes a final selection of twenty-five or so. The articles come from a wide range of sources — magazines, newspapers, and Web sites. Since I live in a small town in a remote part of New Mexico, I depend heavily on submissions from editors and writers. All submissions must follow a few ground rules. For next year's selection, they must have been published in the United States during the 2005 calendar year and must be nonfiction articles about science or the environment. Essays are also welcome. Please include either the entire publication, a tear sheet, or a high-quality photocopy of the original article that clearly shows the author's name, the publication date, and the name of the publication. All submissions must reach me by early January 2006. The best way for publications to guarantee that all their articles are considered for inclusion in the anthology is to place me on their subscription list. Subscriptions and submissions can be sent to Tim Folger, The Best American Science and Nature Writing, 3617 Zia Drive, Gallup, New Mexico 87301. And if any readers would like to discuss or comment on this year's collection, please visit www.timfolger.net.

It has been a pleasure to work with Jonathan Weiner. I highly recommend his book *The Beak of the Finch,* which won a Pulitzer Prize in 1995. As always, I'm exceedingly grateful to my editors at Houghton Mifflin, Deanne Urmy and Ryan Mann, for their patience and for knocking me off my soapbox when necessary, and to Anne Nolan, my beauteous wife of nine months, for picking me up and dusting me off.

TIM FOLGER

Introduction

MY FATHER was always drawing pictures of atoms on his paper napkins at dinner. He told my brother and me that the napkin, his pen, and his hand were all just atoms and empty space — more atoms than stars in the sky, but too small to see.

One summer evening after dinner, I fell out of the biggest maple tree in our back yard in Hillsdale, New Jersey, and clonked my head against the trunk. Then I sat under the tree with a feeling of destiny: I could see them. Someday, I decided woozily, I'd make a trip to Columbia University, where my father lectured about engineering and atomic theory. I'd stand at his blackboard (from which he used to bring us pieces of chalk), and I'd tell the world what it feels like when you see atoms shoot through empty space right before your eyes.

That was my first moment as a science writer. It was also one of the first moments in science writing, because the ancient Roman poet Lucretius spent much of his epic poem *The Way Things Are* trying to help his friends, fellow Romans, and countrymen imagine what it might be like to see atoms, a job that he did at least as well as anyone has done since. Lucretius also wrote brilliantly about life and death, ecology and evolution, religion and rationality, human perception and optical illusions; but his favorite subject was atomic theory.

In one famous passage, the poet told his readers to sit in a dark room, let the sunlight stream in through the slats in the shutters, and watch the dust motes in the sunbeams. Every bit of matter is that turbulent, he said; even if it looks as solid as marble, or our own living flesh, its atoms are perpetually whirling like dust in a room or sparks above a bonfire. Sometimes the atoms crash into each other. Sometimes they stick together to form a strand — a strand that may grow many atoms long, but is still much too small for us to see. And each of those strands (which we would now call molecules) gets hit on all sides by hundreds of atoms pounding into it. That's why the dust motes dance in the air, Lucretius wrote. They dance because so many invisibly small atoms and molecules are always pummeling each bit of dust, though our human eyes cannot see "what urge compels the dancing."

Lucretius, the best science writer of all time, was born in the year 100 BCE, the same year as Julius Caesar. He wrote that passage about dust motes on papyrus around 45 BCE. Physicists no longer believe that atoms swoop around in a block of marble as freely as dust in a breeze. But they do believe that the atoms and molecules in the air or in a glass of water are perpetually crashing into each other. Albert Einstein proved the point in a paper that appeared one hundred years ago, in 1905, which the world remembers as Einstein's "wonder year." He calculated precisely how much a bit of grit would tend to wander and meander in a drop of water if the atomic hypothesis was correct. His calculations matched perfectly with a French experimentalist's observations through a microscope. That paper of Einstein's finally convinced physicists that the atomic theory is a powerful way to look at the universe. (So did his formula $E = mc^2$.) And forty years later, of course, with Hiroshima and Nagasaki, the whole world understood the power of the atomic theory.

A surprising amount of Lucretius' poem looks spectacularly right today, apocalyptically correct. But poor Lucretius might be better remembered if he had written his epic about heroes, gods, and goddesses instead of atoms. Science writing is usually seen as a world apart even though its subjects surround us, fascinate us, and terrify us, even though at their best all of the arts and sciences share the same subject, which is the way things are.

So we are fortunate that so many fine science writers are working today. I am honored to have been invited to introduce *The Best American Science and Nature Writing 2005*. In spite of its title, of course, this anthology is only a small sampling of the best that appeared last year in the United States. There were enough good articles to fill this book a few times over. It's been a pleasure to work with the series editor, Tim Folger, and Deanne Urmy at Houghton Mifflin, sifting through hundreds of articles that appeared on all facets of science and choosing twenty-five of the strongest.

We are learning so much now about the atoms and molecules in living flesh — the field of study known as molecular biology — that the life sciences are beginning to merge with the physical sciences, bringing prospects of new kinds of healing and new kinds of disaster. These discoveries and what we will do with them present some of the biggest themes for any writer in the first years of the third millennium. With this new science, a life science based on atoms, we can hope not only for new cures but also for new evolutionary strides, new genetic enhancements. How do we balance our yearning for cures and enhancements with the chance that we may do ourselves and our species real harm?

In "Getting in Nature's Way," the fine writer and surgeon Sherwin B. Nuland quotes Montaigne, who warned us that we should not meddle too much with nature, because "she knows her business better than we do." And Jenny Everett gives us a poignant case study of the enhancement question and the pain it can bring to patients and their families in her essay "My Little Brother on Drugs." Everett's kid brother, Alex, was nine when he got his first shot of a synthetic growth hormone called Humatrope. Alex's doctors estimate that he might have grown to five feet six without the drug. They can't say how much taller he will grow with it. So his big sister wonders whether it is worthwhile to submit Alex to the pain and discipline of the daily shots and the medicalizing of a short, healthy, happy kid. And besides giving Alex a chance to grow a few inches taller, what else will Humatrope do? The growth hormone is known to stimulate the body to make a protein called IGF-1. Some researchers think IGF-1 may turn out to be a central regulator of human aging. So while his doctors try to enhance Alex's height, they may also be helping to decide, in ways that nobody can predict today, how long or short a time he will have to live.

Nothing at the edge of science and medicine stirs more passion today than the debate over stem cells. James McManus gives us both a political and a personal view of the subject in his polemic "Please Stand By While the Age of Miracles Is Briefly Suspended." His daughter is diabetic, and he feels that President Bush is ruining medicine's chances of saving her. And Connie Bruck gives us a valuable look at the politics of stem cells in "Hollywood Science." California has been a microcosm of the national debate over stem cell policy, and the debate will be analyzed for years, so Bruck's report from the Western Front is likely to be read and reread by historians and politicians. The debate is so polarized that it is hard to see the way things are: how much Hollywood there is in stem cells and how much real promise. Bruck quotes one scientist who campaigned in favor of California's stem cell initiative: "Maybe every hundred years we have one major milestone in medical research"; and stem cells are it. Bruck also quotes a doctor on the other side, who argues that stem cell research will lead to the cloning of human beings. "This is what we would call a clone-and-kill bill! It will make California the mecca of cloning and irresponsible medicine . . . and keep us in budgetary crisis for twenty-five years!"

Well, the Age of Miracles is here now, at least in California, and it will be fascinating to watch how it plays out. The real test will come with the pressure to do something spectacular — a Hollywood spectacular — in the next few years. Watch the pages of the next *Best American Science and Nature Writing*.

Our attitude toward medicine has always been contradictory and passionate, and nowhere more so than in the realm of alternative medicines. In "Miracle in a Bottle," Michael Specter gives us a look at the old nostrum companies in their new bottles, "an extremely irregular business," as he writes, but an extremely ancient one too. The claims made by these companies are often implausible or impossible, but most consumers find it hard to sort fact from fiction. Pseudoscience surrounds us. The editors of *Popular Science* asked William Speed Weed to note down every so-called scientific claim he ran across in the course of an ordinary day. He tells us how the day went in his essay "106 Science Claims and a Truckful of Baloney."

In the science of psychology it is even harder to see the way things are. I've included some contrarian and fascinating articles about a few of the central tenets or practices of psychology. In "Out, Damned Blot," Frederick Crews reviews a skeptical book about the Rorschach inkblot test. Crews argues that the test is as random and arbitrary as the blots themselves. In "Personality Plus," Malcolm Gladwell argues that the Myers-Briggs and the TAT personality tests are just as arbitrary. In "The Grief Industry," Jerome Groopman suggests that one of the most popular practices in current psychotherapy, crisis counseling, may be misguided or even counterproductive. (And see Malcolm Gladwell's "Getting Over It.")

Many psychologists are now turning to machines that can measure brain activity, and in the process they feel they are at last turning psychology into what we like to call a hard science — a science based on bedrock, ultimately on atoms. In his essay "Whose Life Would You Save?" Carl Zimmer visits one of the leading laboratories of this kind, at Princeton, and watches as a philosopher tries to sort out the difference between one kind of morality and another, based on which parts of the brain light up.

This planet still has undiscovered places as little known as human nature. Robert Kunzig takes us to one of them in a *Discover* cover story, "20,000 Microbes Under the Sea." Our interest in Earth's sunlit surface is parochial. The bottom of the sea turns out to hold nearly a third of all the life on the planet. It may also have the power to transform life on Earth through greenhouse gases, a sort of planetary belch of methane, giving at least one scientist a nightmare vision of a "postapocalyptic greenhouse."

And in an undersea explorer's horror story, Jeffrey M. O'Brien introduces us to Bill Stone, who invents sea-diving gear so novel and useful that NASA is one of his biggest customers. We watch the inventor dive into a cave 4,500 feet down beneath Oaxaca, Mexico, to recover the body of a friend who has died testing his invention.

In "Stumbling into Space," Timothy Ferris scathingly reviews the blunders that led to the *Columbia* disaster. Like the explosion of the *Challenger*, this was an accident, but it was also the product of bad judgment and a demoralized and confused bureaucracy. In "The X Prize," Ian Parker shows us the can-do spirit of pioneers outside the bureaucracy. And in a plain, blunt essay, "A Two-Planet Species?" William Langewiesche reminds us of what is at stake here. Unless we learn to live on more than one world someday, we are unlikely to survive. "Compared with the scale of such an ambition," he writes, "a pause of a few decades now to rethink and rebuild will seem like nothing at all." Jared Diamond offers a cautionary tale in his essay "Twilight at Easter," which was a prelude to *Collapse,* his bestseller on the same subject. When the Easter Islanders cut down all their trees, every last species became extinct, and then the islanders died too. "Polynesian Easter Island was as isolated in the Pacific Ocean as the Earth is today in space," Diamond writes. "When the Easter Islanders got into difficulties, there was nowhere to which they could flee or to which they could turn for help; nor shall we modern Earthlings have recourse elsewhere if our troubles increase."

In "Crossing the Red Line," Bill McKibben reviews the current state of our planet in one of its most vital indicators, its temperature. When he first began writing about global warming in *The End of Nature,* a writer could actually keep up with the literature on the subject. Today it is an explosion. McKibben agrees with James Speth, one of the authors whose books he is reviewing here, that our best hope lies in voluntary simplicity. McKibben writes, "It is the true measure of our desperate position that the frail hopes expressed by Speth may turn out to suggest the most solid and practical advice anyone can give."

Strangely, as our impact on the planet grows bigger and bigger, some of our most interesting science grows smaller and smaller, descending into molecular biology and electronics. Two writers, Ellen Ullman and Jennifer Kahn, look at some of the ways the computer revolution is transforming our culture. Ullman, who used to write code for a living, worries that the robots are taking over — not the way science-fiction robots take over but through our involuntary and obsessive attempts to make our lives more and more like the ideal computer, "compact, elegant, error-free." While we keep trying to make our computers more like us at the interface, with avatars, we are meeting the computers more than halfway by trying to make ourselves more like them: "Fast, efficient, untiring, correct, standardized, organized: the virtues we humans strive for but forever fail to achieve, the reasons we invented our helpmate, the machine."

Jennifer Kahn writes, in a profile in *Wired,* of a notorious hacker, Adrian Lamo, who is known as a "grayhat" hacker, someone who probes security systems and then lets major corporations know that he has found a way in. "Grayhats see themselves as Internet Zorros

— high-minded vigilantes," she writes. She finds herself drawn to the grunge glamour and mystique of the hacker's world. But not every corporation sees them the way they wish to be seen. When Lamo hacks into the *New York Times,* the Gray Hat gets in trouble with the Gray Lady of journalism.

Cliff Stoll, who once played a role in catching a hacker, a story he told in his bestseller *The Cuckoo's Egg,* feels nostalgic for one old culture that the computer has erased completely. In "The Curious History of the First Pocket Calculator," he tells the long-forgotten story of the Curta, a machine so beautifully made that it "purrs as you calculate." This extraordinary gadget saved the life of its inventor, who designed it when he was a prisoner of the Nazis in Buchenwald.

The year 2004, an election year, was consumed with talk of faith, and two well-known science writers weigh in to defend the country's agnostics and atheists. Natalie Angier notes in her essay "My God Problem — and Theirs" that many scientists she knows pretend to be more religious than they feel — just as many politicians do. They depend on taxpayers' money and they don't want to be seen as "irreligious, a prionic lifeform bent on destroying the most sacred heifer in America." John Horgan writes in the same spirit in a *New York Times* op-ed, "Keeping the Faith in My Doubt." He notes that the number of people in this country with no religious affiliation at all is now approaching 40 million. "That is twice the number of Muslims, Jews, Buddhists, Hindus, and Episcopalians combined." This puts them in the same camp or party with Lucretius, who argued passionately that science opens the world to us whereas religion closes it. The causes of things are hidden from our eye, Lucretius says, and that is why human beings believe in gods. But once we stop looking for reality there, we find "an open Way / To Nature's Secrets, and we walk in Day." He writes, "So, little by little, time brings out each thing into view, and reason raises it up into the shores of light."

Oliver Sacks shows us some of the best of this kind of thinking — bringing together the old and new — in his lovely essay "In the River of Consciousness," in which he reviews different kinds of evidence for what consciousness is — what quality it has. We experience it as a flow, and yet in special circumstances it is broken up into bits, like frames of a film slowed way, way down. Is perception always like that without our knowing it? Sacks concludes that consciousness is atomized into moments, "but," as he writes, "moments of an essentially personal kind."

With so much opening up to our view, both good and bad, this is an extraordinary moment in science, and there is much to write about. Many of the new things we are seeing now will matter to readers a thousand years from now, if we can only see them clearly enough and record them with passion.

And that is what it is like to see atoms.

JONATHAN WEINER

*The Best American Science
and Nature Writing 2005*

NATALIE ANGIER

My God Problem — and Theirs

FROM *The American Scholar*

IN THE COURSE of reporting a book on the scientific canon and pestering hundreds of researchers at the nation's great universities about what they see as the essential vitamins and minerals of literacy in their particular disciplines, I have been hammered into a kind of twinkle-eyed cartoon coma by one recurring message. Whether they are biologists, geologists, physicists, chemists, astronomers, or engineers, virtually all my sources topped their list of what they wish people understood about science with a plug for Darwin's dandy idea.

Would you please tell the public, they implored, that evolution is for real? Would you please explain that the evidence for it is overwhelming, and that an appreciation of evolution serves as the bedrock of our understanding of all life on this planet?

In other words, the scientists wanted me to do my bit to help fix the terrible little statistic they keep hearing about, the one indicating that many more Americans believe in angels, devils, and poltergeists than believe in evolution. According to recent polls, about 82 percent are convinced of the reality of heaven (and 63 percent think they're headed there after death); 51 percent believe in ghosts; but only 28 percent are swayed by the theory of evolution.

Scientists think this is terrible, the public's bizarre underappreciation of one of science's great and unshakable discoveries — how we and all we see came to be — and they're right. Yet I can't help feeling tetchy about the limits most of them put on their complaints. You see, they want to augment this particular figure — the number of people who believe in evolution — without bothering to confront a few other salient statistics that pollsters have revealed about America's religious cosmogony. Few scientists, for example, worry about the 77 percent of Americans who insist that Jesus was born to a virgin, an act of parthenogenesis that defies everything we know about mammalian genetics and reproduction. Nor do the researchers wring their hands over the 80 percent who believe in the resurrection of Jesus, the laws of thermodynamics be damned.

No, most scientists are not interested in taking on any of the mighty cornerstones of Christianity. They complain about irrational thinking, they despise creationist "science," they roll their eyes over America's infatuation with astrology, telekinesis, spoon bending, reincarnation, and UFOs, but toward the bulk of the magic acts that have won the imprimatur of inclusion in the Bible, they are tolerant, respectful, big of tent. Indeed, many are quick to point out that the Catholic Church has endorsed the theory of evolution and that it sees no conflict between a belief in God and the divinity of Jesus and the notion of evolution by natural selection. If the pope is buying, the reason for most Americans' resistance to evolution must have less to do with religion than with a lousy advertising campaign.

So, on the issue of mainstream monotheistic religions and the irrationality behind many of religion's core tenets, scientists often set aside their skewers, their snark, and their impatient demand for proof, and instead don the calming cardigan of a kiddie-show host on public TV. They reassure the public that religion and science are not at odds with one another, but rather that they represent separate "magisteria," in the words of the formerly alive and even more formerly scrappy Stephen Jay Gould. Nobody is going to ask people to give up their faith, their belief in an everlasting soul accompanied by an immortal memory of every soccer game their kids won, every moment they spent playing fetch with the dog. Nobody is going to mock you for your religious beliefs. Well, we might if you base your life decisions on the advice of a Ouija board; but if you want to believe that someday you'll be seated at a celestial banquet with your long-dead father to your right and Jane Austen to your left — and that she'll want to talk to you for another hundred million years or more — that's your private reliquary, and we're not here to jimmy the lock.

Consider the very different treatments accorded two questions presented to Cornell University's "Ask an Astronomer" Web site. To the query "Do most astronomers believe in God, based on the available evidence?" the astronomer Dave Rothstein replies that, in his opinion, "modern science leaves plenty of room for the existence of God . . . places where people who do believe in God can fit their beliefs in the scientific framework without creating any contradictions." He cites the Big Bang as offering solace to those who want to believe in a Genesis equivalent, and the probabilistic realms of quantum mechanics as raising the possibility of "God intervening every time a measurement occurs," before concluding that ultimately science can never prove or disprove the existence of a god, and religious belief doesn't — and shouldn't — "have anything to do with scientific reasoning."

How much less velveteen is the response to the reader asking whether astronomers believe in astrology. "No, astronomers do not believe in astrology," snarls Dave Kornreich. "It is considered to be a ludicrous scam. There is no evidence that it works, and plenty of evidence to the contrary." Dr. Kornreich ends his dismissal with the assertion that in science "one does not need a reason not to believe in something." Skepticism is "the default position" and "one requires proof if one is to be convinced of something's existence."

In other words, for horoscope fans, the burden of proof is entirely on them, the poor gullible gits; while for the multitudes who believe that, in one way or another, a divine intelligence guides the path of every leaping lepton, there is no demand for evidence, no skepticism to surmount, no need to worry. You, the religious believer, may well find subtle support for your faith in recent discoveries — that is, if you're willing to upgrade your metaphors and definitions as the latest data demand, seek out new niches of ignorance or ambiguity to fill with the goose down of faith, and accept that, certain passages of the Old Testament notwithstanding, the world is very old, not everything in nature was made in a week, and (can you turn up the mike here, please?) Evolution Happens.

And if you don't find substantiation for your preferred divinity or your most cherished rendering of the afterlife somewhere in the sprawling emporium of science, that's fine, too. No need to lose faith when you were looking in the wrong place to begin with. Science can't tell you whether God exists or where you go when you die. Science cannot definitively rule out the heaven option, with its helium balloons and Breck hair for all. Science in no way wants to be associated with terrifying thoughts, like the possibility that the peri-century of consciousness granted you by the convoluted, gelatinous, and transient organ in your skull just may be the whole story of you-dom. Science isn't arrogant. Science trades in the observable universe and testable hypotheses. Religion gets the midnight panic fêtes. But you've heard about evolution, right?

So why is it that most scientists avoid criticizing religion even as they decry the supernatural mindset? For starters, some researchers are themselves traditionally devout, keeping a kosher kitchen or taking communion each Sunday. I admit I'm surprised whenever I encounter a religious scientist. How can a bench-hazed Ph.D., who might of an afternoon deftly purée a colleague's Power-Point presentation on the nematode genome into so much fish chow, then go home, read a two-thousand-year-old chronicle, riddled with internal contradictions, of a meta-Nobel discovery like "Resurrection from the Dead," and say, gee, that sounds convincing? Doesn't the good doctorate wonder what the control group looked like?

Scientists, however, *are* a far less religious lot than the American population, and the higher you go on the cerebro-magisterium, the greater the proportion of atheists, agnostics, and assorted other paganites. According to a 1998 survey published in *Nature,* only 7 percent of members of the prestigious National Academy of Sciences professed a belief in a "personal God." (Interestingly, a slightly higher number, 7.9 percent, claimed to believe in "personal immortality," which may say as much about the robustness of the scientific ego as about anything else.) In other words, more than 90 percent of our elite scientists are unlikely to pray for divine favoritism no matter how badly they want to beat a competitor to publication. Yet only a flaskful of the faithless have put their nonbelief on record or publicly criticized religion, the notable and voluble exceptions being Richard Dawkins of Oxford University and Daniel Dennett of Tufts University. Nor have Dawkins and Dennett earned much good will among their colleagues for their anticlerical views; one astronomer I spoke with said of Dawkins, "He's a really fine parish preacher of the fire-and-brimstone school, isn't he?"

Even a recent D&D-driven campaign to spruce up the image of atheism by coopting the word "bright" — to indicate, as Dawkins put it in a newspaper piece, "a person whose worldview is free of supernatural and mystical elements" — has yet to gain much candlepower. Admittedly, the new lingo has problems. To describe oneself at a cocktail party as "a bright," however saucily and sunnily the phrase is delivered, makes one sound like a member of a Rajneeshee-style cult, not exactly the image a heathen wants to project.

But I doubt that semantics is what keeps most scientists quiet about religion. Instead, it's probably something closer to that trusty old limbic reflex called "an instinct for self-preservation." For centuries science has survived quite nicely by cultivating an image of reserve and objectivity, of being above religion, politics, business, table manners. Scientists want to be left alone to do their work, dazzle their peers, and hire grad students to wash the glassware. When it comes to extramural combat, scientists choose their crusades cautiously. Going after Uri Geller or the Raelians is risk-free entertainment, easier than making fun of the sociology department. Battling the creationist camp has been a much harder and nastier fight, but those scientists who have taken it on feel they have a direct stake in the debate and are entitled to wage it, since the creationists, and more recently the promoters of "intelligent design" theory, claim to be as scientific in their methodology as are the scientists.

But when a teenager named Darrell Lambert was chucked out of the Boy Scouts for being an atheist, scientists suddenly remembered all those gels they had to run and dark matter they had to chase, and they kept quiet. Lambert had explained the reason why, despite a childhood spent in Bible classes and church youth groups, he had become an atheist. He took biology in ninth grade, and rather than devoting himself to studying the bra outline of the girl sitting in front of him, he actually learned some biology. And what he learned in biology persuaded him that the Bible was full of . . . short stories. Some good, some inspiring, some even racy, but fiction nonetheless. For his incisive, reasoned, scientific look at life, and for refusing to cook the data and simply lie to the Boy Scouts about his thoughts on God — as some advised him to do — Darrell Lambert should have earned a standing ovation from the entire scientific community. Instead, he had to settle for an interview with Connie Chung, right after a report on the Gambino family.

Scientists have ample cause to feel they must avoid being viewed as irreligious, a prionic lifeform bent on destroying the most sacred heifer in America. After all, academic researchers graze on taxpayer pastures. If they pay the slightest attention to the news, they've surely noticed the escalating readiness of conservative politicians and an array of highly motivated religious organizations to interfere with the nation's scientific enterprise — altering the consumer information Web site at the National Cancer Institute to make abortion look like a cause of breast cancer, which it is not, or stuffing scientific advisory panels with antiabortion "faith healers." Recently, an obscure little club called the Traditional Values Coalition began combing through descriptions of projects supported by the National Institutes of Health and complaining to sympathetic congressmen about those they deemed morally "rotten," most of them studies of sexual behavior and AIDS prevention. The congressmen in turn launched a series of hearings, calling in institute officials to inquire who in the Cotton-pickin' name of Mather cares about the perversions of Native American homosexuals, to which the researchers replied, um, the studies *were* approved by a panel of scientific experts, and, gee, the Native American community has been underserved and it is having a real problem with AIDS these days. Thus far, the projects have escaped being nullified, but the raw display of pious dentition must surely give fright to even the most rakishly freethinking and comfortably tenured professor. It's one thing to monkey with descriptions of Darwinism in a high school textbook . . . but to threaten to take away a peer-reviewed grant! That Dan Dennett; he is something of a pompous leaf-blower, isn't he?

Yet the result of wincing and capitulating is a fresh round of whacks. Now it's not enough for presidential aspirants to make passing reference to their "faith." Now a reporter from *Newsweek* sees it as his privilege, if not his duty, to demand of Howard Dean, "Do you see Jesus Christ as the son of God and believe in him as the route to salvation and eternal life?" In my personal fairy tale, Dean, who as a doctor fits somewhere in the phylum Scientificus, might have boomed, "Well, with his views on camels and rich people, he sure wouldn't vote Republican!" or maybe, "No, but I hear he has a

Mel Gibson complex." Dr. Dean might have talked about patients of his who suffered strokes and lost the very fabric of themselves, and how he has seen the centrality of the brain to the sense of being an individual. He might have expressed doubts that the self survives the brain, but, oh yes, life goes on, life is bigger, stronger, and better endowed than any Bush in a jumpsuit, and we are part of the wild, tumbling river of life, our molecules were the molecules of dinosaurs, and before that of stars, and this is not Bulfinch mythology, this is corroborated reality.

Alas for my phantasm of fact, Howard Dean, M.D., had no choice but to chime, oh yes, he certainly sees Jesus as the son of God, though he at least dodged the eternal life clause with a humble mumble about his salvation not being up to him.

I may be an atheist, and I may be impressed that, through the stepwise rigor of science, its Spockian eyebrow of doubt always cocked, we have learned so much about the universe. Yet I recognize that, from there to here, and here to there, funny things are *everywhere.* Why is there so much dark matter and dark energy in the great Out There, and why couldn't cosmologists have given them different enough names so I could keep them straight? Why is there something rather than nothing, and why is so much of it on my desk? Not to mention the abiding mysteries of e-mail, like why I get exponentially more spam every day, nine-tenths of it invitations to enlarge an appendage I don't have.

I recognize that science doesn't have all the answers and doesn't pretend to, and that's one of the things I love about it. But it has a pretty good notion of what's probable or possible, and virgin births and carpenter rebirths just aren't on the list. Is there a divine intelligence, separate from the universe but somehow in charge of the universe, either in its inception or in twiddling its parameters? No evidence. Is the universe itself God? Is the universe aware of itself? We're here. We're aware. Does that make us God? Will my daughter have to attend a Quaker Friends school now?

I don't believe in life after death, but I'd like to believe in life before death. I'd like to think that one of these days we'll leave superstition and delusional thinking and Jerry Falwell behind. Scientists would like that, too. But for now, they like their grants even more.

CONNIE BRUCK

Hollywood Science

FROM *The New Yorker*

ADDRESSING A state legislative committee hearing held in San Diego on September 15, a young research scientist named Hans Keirstead declared, "Maybe every hundred years we have one major milestone in medical research." This was such a moment, he said. The committee, chaired by State Senator Deborah Ortiz, was hearing testimony on Proposition 71, a ballot initiative that would result in the state's investing about $3 billion in stem cell research over the next decade. Stem cells are undifferentiated cells that have the capacity to renew and develop into other types of cells. Prompted by research restrictions on embryonic stem cells imposed by the Bush administration and by threats from its allies in Congress, the initiative constitutes a breathtakingly large debt for California's taxpayers, particularly in light of the state's continuing budgetary crisis and cuts in education, health care, and public safety. According to a Field poll published in August, 45 percent of California voters favored the initiative and 42 percent opposed it. Keirstead's real audience at this hearing, therefore, was not its presiding officials but the rows and rows of public spectators in the big, packed hearing room and the cameras from a half-dozen TV networks.

Critics of embryonic stem cell research often accuse its supporters of promising imminent cures that are at best years away, if feasible at all. Keirstead, whose tanned skin, wavy, sun-streaked hair, and stylish suit hardly suggested a specialist in the study of spinal-cord injuries or days spent in subterranean quarters at the University of California at Irvine with hundreds of caged animals, showed a video in which he attempted to counter such charges. In one scene, a small gray rat struggles to move forward, dragging itself by its front paws. The next scene, Keirstead explained, would show a rat that had suffered the same spinal-cord injury but had received a transplantation of human embryonic cells that had been directed to differentiate into oligodendrocytes, a particular type of brain cell. As Keirstead had told me earlier, the very quality that epitomizes the promise of embryonic stem cells — their amazing plasticity — can also be their great failing ("Just stuff stem cells into a brain lesion with a scar, and the cells will make scar tissue"). He had succeeded, he said, in promoting the differentiation of the stem cells into precisely what he wanted — oligodendrocytes, in high purity. And, in the next clip, the rat appears — walking! Its tail lifted! Keirstead predicted that sometime in 2006 there would be a clinical study involving humans with spinal-cord injuries.

Scientists have studied embryonic stem cells from mice, in culture, for decades, but it was only six years ago that James Thomson, a researcher at the University of Wisconsin, was able to isolate and maintain human embryonic stem cells in a similar way. Some of the hope currently fixed on these cells is based on the far more extensive history of adult stem cells, which are found in fetuses, umbilical-cord blood, children, and adults. Hematopoietic stem cells, for example — the blood-forming cells found mainly in the bone marrow, and discovered about forty years ago — were the first stem cells to be used successfully in blood-disorder and cancer therapies, such as bone marrow transplants. Some researchers believe they may have the potential for more wide-ranging treatments. But adult stem cells are present only in minute quantities in the body, they are difficult to isolate successfully, and — thus far — it has been nearly impossible to grow them outside the body.

Embryonic stem cells, however, are capable of becoming almost any type of cell or tissue, and they are self-renewing, whether in the body or in the laboratory. They are found in embryos about five days after fertilization, when the embryo becomes a shell called a blastocyst, comprising about two hundred cells; if removed from the blastocyst and placed in a culture dish under the right conditions, these cells become a stem cell line, which continues to propagate indefinitely in the laboratory, and — scientists believe — can be directed to produce large amounts of tissue. Thus, embryonic stem cells appear to have unique potential to lend themselves to a broad array of disease therapies, creating new healthy tissue to replace damaged or dead tissue, such as pancreatic islet cells for type 1 diabetes, or various neuronal cells for spinal-cord injuries, ALS, and Parkinson's disease.

But the very quality that embodies such promise — the capacity of the stem cells to form any tissue — derives from these cells' representing, if only as a microscopic dot, life's origins. And removing the cells from the blastocyst to create a self-renewing stem line destroys the embryo. The Catholic Church and some other religious institutions, therefore, oppose embryonic stem cell research. And in August 2001, President Bush announced that he would allow federal funding only for research on embryonic stem cell lines already cultivated in laboratories, and he prohibited funding for the development of new lines, which would require the destruction of additional embryos.

This fractious conflation of science, religion, and politics had brought a dramatic cast of characters to the San Diego hearing room. The president of the "No on Prop 71" campaign, an emotive physician named Vincent Fortanasce, reviled embryonic stem cell research as something that would cause far more disease than it cured, since under certain conditions embryonic stem cells can give rise to tumors. He also swore that its technology would lead inevitably to the cloning of human beings. "This is what we would call a clone-and-kill bill!" he exclaimed. "It will make California the mecca of cloning and irresponsible medicine . . . and keep us in budgetary crisis for twenty-five years!"

Although Fortanasce is a devout Catholic, he did not make a pro-life argument. He and other opponents of Prop 71 emphasize that their coalition includes fiscal conservatives and pro-choice feminists; their lead lobbying group is called Doctors, Patients, and Taxpayers for Fiscal Responsibility. But I had interviewed Dr. Fortanasce a couple of weeks earlier, and he told me, "If you want to know who is organizing this, it is the Catholic Conference. I wouldn't want you to find that out later and feel I'd misled you."

Fortanasce was followed by the feminist Diane Beeson, a sociology professor, who said she fears the initiative will result in poor women being encouraged to sell their eggs for research, and that the technology will be used not only for disease therapies but also for cosmetic surgery and "made-to-order children." Every witness, pro or con, had strong convictions. Mothers pleaded for the proposition's passage, as their brain-damaged children sat beside them in wheelchairs. A cancer survivor said that the initiative ought to emphasize research on adult stem cells, which had helped him, rather than embryonic stem cells. A woman representing the Snowflakes Embryo Adoption Program said that the excess embryos at in vitro fertilization clinics — most of which are slated for destruction if not given to research — ought to be adopted instead. A man from an organization called Biblical Family Advocates said that even if Prop 71 resulted in cures for diseases, neither he nor any member of his family would accept them. "Some say it's just a blob of cells," he declared. "Jesus made it very clear, when you did it to the least of these, you did it unto me." The hearing was by turns heart-wrenching, provocative, and inspiring, but more than anything else it revealed what an odd, ill-suited process this was for deciding a complex and momentous matter of public policy.

For many decades, decisions about federal funding of scientific research have been made at the National Institutes of Health (NIH), the world's leading medical research institution, which awards more than forty thousand research grants each year to scientists across the country. An agency within the Department of Health and Human Services, the NIH is beholden to the executive branch and to Congress, which influences its research agenda. This relationship has been particularly salubrious in the past several years; the NIH budget has doubled since 1998, going from $13.6 billion that year to $27.3 billion in 2003. Last year, the NIH spent $190.7 million on human adult stem cell research, which is not controversial, and $24.8 million on embryonic stem cell research.

Since 1996, a congressional ban has prohibited the use of federal funds to create, damage, or destroy human embryos for research. But after Thomson's breakthrough in isolating the human embryonic stem cells, in 1998, the Clinton administration's NIH found a way around the ban. Because stem cells were not embryos, NIH lawyers concluded that the law did not apply. Therefore federal funding could be used for research involving embryonic stem cells but not for creating them (the process that destroys the embryo would have to be privately funded). A group was convened to pro-

duce a set of ethical guidelines for the research, and Paul Berg, a
Nobel laureate, served as a consultant. Berg had been in the vortex
of politics and science before. In 1974, when he and his colleagues
discovered how to engineer recombinant DNA, it provoked fears of
rampant super-microbes that might be created by some errant laboratory; Berg was instrumental in drawing up the strict regulations
that have allowed such DNA research to flourish.

While the risks of recombinant DNA research had seemed frightening at the time, Berg told me recently, the benefits were still theoretical. "There was nothing that anybody had made yet — nobody
had made insulin, nobody had made erythropoietin, nobody had
made interferons — but we all understood that the promise existed," Berg said. "The NIH jumped in with both feet, hands, head,
everything, and funded this research and enabled it to move forward aggressively — that's why we got all the biotechnology benefits of insulin within two or three years.

"Nonscientists find it very difficult to grasp that once you initiate
a technology breakthrough — particularly if it's fundamental and
provides what we call a platform technology — the world out there
is waiting for a chance to exploit that. You let loose a new idea, a
new technology, everybody runs with it — and what you find is
there are a lot of very smart people out there. And they find things
that the initial discoverers never conceived of."

By the time the Clinton administration had finished building the
foundation for the NIH to fund embryonic stem cell research, the
2000 presidential campaign season had begun, and "there was too
much anxiety about the election to move forward," a former NIH
official said. During the campaign, Bush said he was opposed to
embryonic stem cell research; once he was elected, many in the
scientific and patient-advocacy communities expected him to ban
funding for it altogether. So when he announced, in August 2001,
that federal funding could continue on the already existing stem
cell lines, which would not entail the creation and destruction of
embryos, Berg and his colleagues were somewhat relieved. Private
and state funding of stem cell research could continue, and federal
funding could, too, on what Bush said were "more than sixty" existing lines.

But the scientists were also perplexed. They knew of only about
five lines, at the University of Wisconsin. "What turns out is that

most are useless," Berg said. "Most were not cell lines. A cell line is something that is self-renewing, that you can perpetuate indefinitely. These were things where people had just taken a blastocyst, cracked it open, harvested some cells, put them in the deep freeze. The number was reached almost overnight." Today, James Battey, who has been head of the Stem Cell Task Force at the NIH since mid-2002, says that there are twenty-two lines available for government-approved research. Asked how Bush's "more than sixty" was arrived at, Battey said, "I have no idea." He also acknowledged that the lines have other shortcomings — for example, they offer little genetic diversity. They have all come from excess embryos at in vitro fertilization clinics, where patients are typically white and affluent.

About a week before President Bush's announcement, a bill that he supported, known as the Weldon bill, after Representative David Weldon, of Florida, passed in the House. A similar bill was introduced in the Senate by Sam Brownback, Republican of Kansas. The bills would ban all human cloning — not only reproductive but also therapeutic. Reproductive cloning, if it could be done successfully, would result in the creation of a cloned human. A 2002 report by the National Academy of Sciences concluded that, based on the high rate of abnormalities in cloned animals, human cloning is both dangerous and likely to fail; therefore, without even reaching broader ethical issues, the panel called for a legally enforceable ban. The same panel, however, came to a different conclusion about somatic-cell nuclear transfer, also known as therapeutic cloning (this, too, was only theoretical until early this year, when it was done successfully in a study in South Korea). In both processes, a nucleus from a cell is inserted into an egg from which the nucleus has been removed, and the egg takes on that cell's genetic characteristics. In reproductive cloning, the egg becomes a blastocyst that would be inserted into a woman's uterus and, in theory, grow to be a baby that would be a near genetic copy of its donor. In therapeutic cloning, the egg becomes a blastocyst that contains the primitive stem cells; those are removed — a process that destroys the blastocyst — and the stem cell lines are created. The characteristics of those lines, of course, depend on the cell nucleus the researchers injected.

If their aim was to improve on the limited, government-approved lines by better representing the genetic diversity of the American public, they might have inserted skin-cell DNA from African Americans, American Indians, Asians, and so forth. If they wanted to facilitate the study of genetic diseases, they would have injected a disease nucleus and then watched how the disease develops in the stem cell line. And if they wanted to create stem cell lines that would be genetically compatible with patients, so that the lines could be used to develop tissues or organs to replace those that were lost or damaged, they would inject a skin-cell nucleus from the patient.

The Weldon and Brownback legislation would make all these activities criminal, with penalties of up to ten years in prison and a $1 million fine. Moreover, an American who goes abroad for a therapy derived from this process could be subject to these penalties on returning to the United States.

In my conversation with Berg, he said, "Congress is saying, 'We're prepared to deny two hundred and ninety million people in this country access to a therapy that could be lifesaving, because we're offended by the technology.'" He paused, as though he hardly knew how to respond to something so outrageous. "It's what I call un-American, to even think in those terms!" He added that the threat of such legislation, combined with the constraints imposed by President Bush, had discouraged young scientists from entering the field.

In an effort to defeat the Brownback bill, two prominent Hollywood couples formed an organization they called Cures Now. Both couples — Lucy Fisher and Doug Wick, and Janet and Jerry Zucker — had children who had been diagnosed with type 1 juvenile diabetes. At the time, Tessa Wick was eight years old and Katie Zucker was eleven. Their parents became active in the Juvenile Diabetes Research Foundation (JDRF), a patient-advocacy group comprising mainly type 1 diabetes families, and they learned that the best hope for a cure lay in stem cell research. In type 1 diabetes, the body attacks its own beta cells, which regulate blood-sugar levels; if those cells could be restored, the disease could be cured.

The JDRF distributes about $100 million annually to research. While it has funded embryonic stem cell research for the past several years, it has yet to fund any research that involves therapeutic cloning. Its chief executive, Peter Van Etten, explained that it is important to many JDRF supporters that the stem cell research is done on excess embryos from in vitro fertilization clinics, which would likely be discarded. But in therapeutic cloning one is essentially creating an embryo — or at least a manipulated egg — for research purposes. The JDRF, which carefully cultivates relations on Capitol Hill, chose not to fight Brownback.

Therapeutic cloning is, indeed, a line that many do not wish to cross. The columnist Charles Krauthammer, who sits on the President's Council on Bioethics, stresses that he does not believe that personhood begins at conception, believes in legalized abortion, and supports stem cell research on leftover embryos from in vitro fertilization clinics. But he opposes creating an embryo for experimental use and destruction. Essentially, his is a slippery-slope argument: "Violate the blastocyst today and every day, and the practice will inure you to violating the fetus or even the infant tomorrow," he wrote in the *New Republic*.

"Everybody said you can't beat Brownback," Janet Zucker recalled, when I spoke with her and her husband in September. "They felt you don't want to turn off the administration when they're taking little steps forward" — by not banning federal funding altogether. "But one step forward is not good enough for us, because our kids have a ticking time clock." "What has us scared to death," Jerry Zucker picked up, "are the down-the-line implications — loss of sight, loss of limbs. As a parent, you can't just wait, hoping other people do something to find a cure for your daughter."

The Zuckers and Fisher and Wick began holding fund-raising dinners at their homes (where Nobel laureates held forth for the movie crowd) and shuttling back and forth to Washington. The fact that Wick's movie *Gladiator* had become a blockbuster and won an Academy Award increased his cachet among some politicians; Zucker, for his part, had codirected *Airplane!* and directed *Ghost*. Wick's father, Charles Wick, was the head of the U.S. Information Agency under the Reagan administration, and the Reagan and Wick families have been close for decades. Nancy Reagan, who had attended one of the scientific briefing dinners at Fisher and Wick's house, called many Republican members of Congress and told them how important she believed embryonic stem cell research was. And Doug's father called Karl Rove and asked him to meet with them. "He was one of our first stops," Fisher said. "He said two things. First, how it would be really bad if you got people's hopes up. And then he said, Well, it would lose this many votes, gain this many. It was just a political calculation. I think he thought we'd go away. But he didn't understand. If you're a parent of a sick child, you're never going away."

Senator Orrin Hatch, Republican of Utah, came out against the Brownback bill, arguing that, as a pro-life advocate, he differentiated between an embryo in a petri dish and a fetus in its mother's womb. With Democrats Dianne Feinstein and Edward Kennedy, Hatch and his Republican colleague Arlen Specter introduced a bill that would ban reproductive cloning but allow therapeutic cloning. In the end, neither their legislation nor Brownback's had the necessary votes.

In September 2002, the California legislature passed a bill allowing therapeutic cloning, and Governor Gray Davis signed it into law. It meant, in effect, that California was a haven for embryonic stem cell research — but it remained unfunded. State Senator Deborah Ortiz, who had sponsored the stem cell measure, next introduced a $1 billion bond measure to fund the research, but it was blocked. She began to consider using the initiative process — California's powerful vehicle for circumventing the legislature and going directly to the people to make public policy — and met with Cures Now in the spring of 2003. When Peter Van Etten, of the JDRF, heard about the initiative idea, he immediately contacted one of his board members, Bob Klein, a real estate developer from Palo Alto who had a child with juvenile diabetes. The previous year, supplemental funding for diabetes research at the NIH had been about to expire, and the only way to get an extension was with the unanimous consent of the 535 members in both houses of Congress. With Klein's help, Van Etten and a few others had obtained $1.5 billion in funding.

Klein, a lean, fit-looking man of fifty-nine, quickly took control of the Prop 71 effort. He decided that the money should be raised by bonds issued in the amount of not $1 billion but $3 billion — a sum that even Irving Weissman, the Stanford University School of Medicine professor and stem cell researcher who is one of the initiative's major backers, acknowledged he at first found "shocking." But Klein argues that $1 billion would not support what he has devised: a ten- to thirteen-year program, moving from basic applied research to clinical trials to therapy development. This is what it will take, he said, "if you want to run a substitute national program." He tried to put it in perspective: The NIH is spending $190 million a year on adult stem cell research that, he said, may benefit victims of eight to twelve diseases. "We're addressing potential benefit to seventy diseases and injuries with two hundred and fifty million dollars a year," he said. Then he took a slightly different tack. "Am I going to tell two-thirds of the diseases that I'm not going to address their issues?"

Having created this towering financial hurdle, Klein tried to design the initiative to meet it. To do so, he relied on the usual California economics: pushing debt obligations off into the future. For at least the first five years, the measure would demand little of the state; the repayment of the bonds' principal would be postponed, and the interest on the debt would be repaid using bond proceeds. After that, however, payments — a total of approximately $6 billion, with principal and interest — would come from the state's general fund. Klein argues that the debt could be repaid by revenues generated by patents, royalties, and license fees from the research; he likes to say that, unlike most bond measures, which finance things like construction of schools and roads, these bonds will give the state an "upside." He predicts that the state's spiraling health care costs will diminish because of the therapies and cures developed; that the California biotech industry will seize the opportunity to become the world leader in this area of research, rather than abdicating its role to the countries that are now investing in it (Britain, South Korea, Singapore, Japan, Sweden, and Israel); and that this, in turn, will result in more jobs and a gain in taxable income. After reviewing these arguments, the state Legislative Analyst's Office concluded, pointedly, "The likelihood and magnitude of these and other potential indirect fiscal effects are unknown."

Klein also crafted the initiative with an eye to preventing any future state or federal intervention. "One of our goals," his mantra goes, "is to take politics out of medicine." The result, however, is

that $3 billion of public money will be entrusted to an entity that is
remarkably autonomous and insular. The initiative's core clauses
are to be incorporated as an amendment to the California Constitution, creating a constitutional right to conduct stem cell research
in California. There is little opportunity for legislative oversight.
Neither the governor nor the legislature can stop the research's annual funding or appropriate the funds for other purposes. A newly
created California Institute for Regenerative Medicine, which is to
award the research grants, would be governed by a twenty-nine-
member board, consisting of representatives of universities, medical research institutions, and patient-advocacy groups; some would
be appointed by various state officials, including the governor, the
treasurer, and the controller.

Ideological opponents have exploited the initiative's structure
by reaching beyond their base with more mainstream arguments.
As Carol Hogan, a spokesperson for the Catholic Conference, said,
"Once we saw the initiative — it makes me think of that Elizabeth
Barrett Browning poem, 'Let me count the ways' — there is so
much to hate about it! So many things, from a public policy point
of view."

Even Ortiz, a strong supporter, emphasized that Prop 71 involved a great deal of public money, and said that she is troubled
by the initiative's lack of allowance for legislative oversight and
greater accountability. How will the deals be struck about the provi-
sions for the state's obtaining royalties? How will decisions be made
about which diseases receive the greatest resources? Ortiz pointed
out that there is a prevalence of type 1 diabetes, for example, in the
families of Prop 71's biggest backers. Will type 1 diabetes receive a
disproportionate amount of research funding? And there is an almost palpable gold-rush excitement among some of the scientists
who have signed on. At one point, the initiative's drafters considered a provision specifying that the scientists making decisions
about grant awards should be from out of state, but it did not make
the final cut. Indeed, the more one ventures into the Prop 71
world, the more one appreciates the role played historically by the
NIH.

The most important of the initiative's twenty-nine board members will be the chairperson, who will manage the agenda, including approval of grants and loans; guide the institute's financing;

deal with the California legislature and the U.S. Congress; lead negotiations for intellectual property agreements; and more. The chairperson's qualifications are enumerated in detail in the initiative. When I pointed out to Klein that these conformed to a remarkable degree with his own résumé, he did not demur. "I have to get back to earning a living," he said. "But if the board believed that for a limited period of time, eighteen to twenty-four months, I should be chairperson, then I would, if I could afford it."

Klein has run this campaign almost as though he were running for office — and, in a way, perhaps he has been. Having set a goal of $20 million for an ad campaign on radio and television, Klein has received substantial contributions from Microsoft's Bill Gates; the eBay cofounders, Pierre and Pam Omidyar; the Amgen founder, William Bowes, Jr.; Senator Jon Corzine, of New Jersey; and others. Klein himself has contributed more than $2 million. He has secured $1 million in funding from the JDRF — the first time that the foundation has contributed to a state-initiative campaign. He has gone to Fresno to speak to farm workers. He has addressed newspapers' editorial board meetings, and chambers of commerce throughout the state. He has helped raise money and won important allies, like State Controller Steve Westly and Treasurer Philip Angelides, both of whom have endorsed the initiative. And he has paid attention to those who will implement the initiative — and who have a vested interest in presenting its best case to the public.

Hans Keirstead, the spinal-cord-injury researcher, is among the scientists Klein has cultivated. Keirstead is also an entrepreneur; he has started two biotech companies and sold one. He has a collaboration with Geron, the California biotech company that funded the Thomson breakthrough at the University of Wisconsin. Geron is exceptionally aggressive about intellectual property rights, but Keirstead said that his lawyers, after long, arduous negotiations, brokered a deal between U.C. Irvine and Geron. "Geron is an intellectual property machine," Keirstead said. "It is the Bill Gates of this world." Geron provides him with training and cells, and made possible $500,000 of the $750,000 his lab receives annually. Still, he believes the passage of Prop 71 would change his life.

He recalled a phone call from Klein. "Bob said, 'You're one of about twelve major people. You just may be benefiting from this bill.' I think he wanted to be sure I was on board," Keirstead told me. He pointed out that the way Klein has structured the initiative — and is selling it — he needs revenues to be generated by the end of the first five years, when the state must begin making payments on the loans. And if Keirstead's experiments go to clinical trials sometime in 2006, they might conceivably produce revenues by the time the state needs them — depending, of course, on what kind of deal the state was able to strike. Now Keirstead walked through the math. Say, $300 million a year, of which perhaps $50 million would go into the construction of research facilities; divide twelve — "OK, even say it's twenty major people, not twelve" — into $250 million. "I could get at least ten million," he concluded. "It would be huge."

Klein also met with Tom Okarma, a former Stanford professor who is now the chief executive officer of Geron. Okarma told me he would not divulge the details of their discussion, because one of the opposition arguments is that the initiative would be a form of corporate welfare — funding with public money what should be done by private industry. "So Geron can't be waving the Prop 71 flag," he said, until it passes. All Geron can do now, Okarma continued, "is to let people in California know that this is not crappy science, that it is within reach, that we are a year and a half from putting these cells into people with spinal-cord injuries" — a reference to Keirstead's experiments. Okarma claimed they are also just a few years away from clinical trials for treatments for heart failure and diabetes — and could move faster if they were not so underfunded. Could Geron get money from the institute created by Prop 71? "Sure we could!" Okarma said.

"We really are on the cusp of something so exciting and important that to have to slalom through the course of threatening legislation and a lethargic NIH is mind-boggling," he continued. He had got a glimpse of the ideology behind the Bush administration policy in 2001 when he was called to testify about embryonic stem cell research before the President's Council on Bioethics; its chair, Dr. Leon Kass, an ethicist from the University of Chicago, now at the American Enterprise Institute, was appointed by Bush that year. In 2002 the council had voted in favor of recommending a four-year moratorium on therapeutic cloning. (A dissenting vote had been cast by Dr. Elizabeth Blackburn, a biochemistry professor at the University of California at San Francisco, who continued to challenge what she thought was the biased nature of council discussions on subjects such as stem cell research. Last February, she was effectively dismissed from the council. She told me that she should have understood from the start that diversity was not desirable; when she was being vetted for the council in 2001, she was asked whom she had voted for in the 2000 election. As an Australian who had not yet become an American citizen, she responded that she had not voted.) Okarma, for his part, said he was "flabbergasted" at the nature of the conversation among certain council members: "The metaphor they used to describe human embryonic stem cells was children who are taken from their home, killed, and their organs distributed for transplantation."

Ideology led to the intrusion of politics into science, Okarma said, and now the science has, predictably, become a "political football," with dubious claims made by both sides. He mentioned Nancy Reagan's remarks, shortly before her husband's death, about embryonic stem cells and Alzheimer's. "We don't know enough about Alzheimer's to even think about how to use embryonic stem cells to cure the disease, let alone to reconstruct the brain," he said. On the other side, Okarma continued, ideological opponents of embryonic stem cell research have propagated claims about the ability of adult stem cells to transdifferentiate, or mature into another type of tissue, as the pluripotent embryonic cells do. The question is indeed controversial, and there are scientists who believe that adult stem cells may be able to transdifferentiate, but Okarma was having none of it. "They say bone marrow cells could become liver, heart, brain — we don't need these immoral embryonic stem cells! It's completely specious. Adult stem cells do not transdifferentiate. You plant a turnip, you get a turnip, not a radish."

Stanford's Irv Weissman makes the same argument. Weissman was the first scientist to identify and isolate adult stem cells in any species, and the first to isolate blood-forming stem cells in humans. (He was the cofounder of two companies that perform research with adult stem cells.) Regarding claims about adult stem cells' ability to transdifferentiate, Weissman said, "I wanted that to be true. I mean, first, we actually did discover blood-forming stem cells, so I feel a little of the glory would rub off on me. And also I have a company that has the patent, and a license from Stanford — I could be the richest person on earth! I had every possible motivation to want that to be true. But it's not. Every one of the claims we tracked down turned out not to be true."

I interviewed Weissman in his small, cramped office, where paper cascades from piles on the desk, chairs, and floor, in Stanford's Beckman Center. Two years ago, he was named the head of a new research center at Stanford — the Institute for Cancer and Stem Cell Biology and Medicine — but it is a "virtual" institute, he explained. He has managed to raise about $15 million to fund it — not enough for a building or for research. If Prop 71 passes, his institute may very well be virtual no longer. (He mentioned, hypothetically, that he might apply for a grant for Stanford for $50 million to build facilities.) He began to describe what he plans to do immediately after a victorious November 2. He will pay a visit to a couple of researchers at Harvard whom he considers the best in the field of stem cell research. "I'm going to say, 'If you come to my new institute at Stanford — a new building, by the way, funded by the state of California — there's going to be thirteen years of funding in your field. Anything that you want to do that would move this field faster toward clinical therapies, you can't do where you are. You'll have great colleagues. You'll be able to do the research, and, by the way, MIT doesn't have a great hospital full of researchers who want to study the diseases that your cell lines will allow them. You'll come here and you'll have collaborators, and, by the way, do you like the winters? Because we don't have those winters.'" Weissman smiled, and his eyes glinted with anticipation. After that, he continued, he will move on to "my friend Harold Varmus's institute" — Sloan-Kettering, in New York. Others in California who also hope to create stem cell institutes funded by the passage of Prop 71 — at the University of California at San Francisco, the University of California at San Diego, the Salk Institute for Biological Science, the California Institute of Technology, the Burnham Institute — will go prospecting, too. "We're going to invade the country for the best, and some will not want to leave, but a lot will. When they leave, those states will be at a competitive disadvantage, and probably only then will the state legislatures and the federal government wake up to realize that California is stealing, for a song, these scientists — and building a whole future."

Compared with Okarma — the chief executive of a public company, responsible to shareholders, which must focus on specific approaches most likely to yield commercial products — Weissman is free to take the long view. He is intent on using nuclear transfer to develop cell lines for every human genetic disorder in an effort to understand how the disorders work and which genes need to be fixed in order to treat and, ultimately, prevent them. "Is this nuclear transfer the same big platform that recombinant DNA was? In my mind, it certainly is, if not more. Diseases we can't approach now, it would enable us to approach. Therapies that we would spend years and years trying to get to, we can probably short-circuit. One thing I know about biomedical science — once you're onto something, once you get the best and the brightest funded to work on it, things move very, very fast." Last January, he met with James Battey, the NIH's Stem Cell Task Force head, and the agency's boss, Elias Zerhouni. "I said, 'Look, what if we make a cell line from ten diabetics'" — using the nuclear transfer, or therapeutic cloning, process. "'Can we at least send these cell lines to the NIH labs, full of people who want to study diabetes?' And they said no way. It is not allowed by the president's executive order.

"California has decided to pick up where the federal government has given up its responsibility to further people's health," Weissman continued. "This is a mini-NIH, doing what the NIH has failed to do."

When I asked Battey about the NIH's rejection of Weissman's proposal, he replied, "Right. We couldn't, because Irv would be creating that cell line with cells created after August 9, 2001. So he could do it with private funds, but we couldn't use federal public funds to study it."

Battey acknowledged that there are currently roadblocks that the NIH cannot get past. "So, to be honest, we're working on the ones where there is something we can do, because there's not a lot of point in fuming about the ones you can't do, right? That's cursing the darkness, as opposed to lighting a candle.

"We're not policymakers. We inform policymakers," he continued quietly. "And the policy, of course, is not made solely based upon scientific input."

Battey mentioned that he had trained at Stanford, in a laboratory next to Weissman's, and a close friend of his had worked for

Weissman. "Oh, I attended their lab meetings, I went to parties at Irv's house, I know Irv Weissman very well," he continued. "He's articulate, brilliant, visionary. Irv is special. He may be the most accomplished hematopoietic stem cell biologist in the world."

And the initiative?

"I'm a very keen follower of it," Battey said. "I think it's very exciting. I must say I'm amazed if the California taxpayer is ready to pony up three billion dollars. But I think such an investment could have a really transforming effect on stem cell research — and it would certainly make California an extremely attractive place to conduct it."

Harold Varmus, the NIH's former head, now at Sloan-Kettering, says he, too, is a strong supporter of the initiative. "I think that the symbolic value of having some state doing it with public money, to embarrass the federal government and try to encourage it to get back in the game, would be very, very important. And California is well equipped to do it. But there really is no substitute for the NIH. And it's a bad precedent, in a way, to say we'll solve the problems of the NIH by doing this, state by state."

The research program Weissman envisages is "very, very attractive," Varmus concluded. But, more than anything, he wants the federal government to change course and create a level playing field across the country for this research. "I don't want to feel that any of my investigators are now going to want to go to California."

Prop 71, of course, owes its very existence to George W. Bush, but that is the best-kept secret of the initiative's campaign. No radio or television ad will mention Bush. "To win, we need Republican votes," one Prop 71 advocate explained. The higher the visibility of stem cell research in the presidential campaign, the more polarizing it becomes — which cuts against Prop 71's campaign message of studious universality. It was, therefore, bad news for the campaign last week when John Kerry brought embryonic stem cell research to the fore, charging that Bush's conduct in this area was consistent with his handling of many other issues. "Now we stand at the edge of the next great frontier, but, instead of leading the way, we're stuck on the sidelines," Kerry said, at a town hall meeting in New Hampshire. "President Bush just doesn't get it. Faced with the facts, he turns away. Time and time again, he's proven that he's stubborn, he's out of touch, he's unwilling to change."

Before Ron Reagan popularized this issue at the Democratic Convention, there was little hint that it might become what Kerry's communications director, Stephanie Cutter, was now calling "the sleeper issue of the campaign." Dr. Jeffrey Bluestone, the director of the Diabetes Center at the University of California at San Francisco, who appears in a Prop 71 television ad, said, "I wonder if it was good for the cause for Ron Reagan to get up at the Democratic Convention. I think it was a double-edged sword — because, early on, embryonic stem cell research was not seen as so political, and now it is."

Ron Reagan told me that after his father died he got a call from John Kerry. "He promised me that if he was elected his first act as president would be to sign an executive order reversing Bush's policy on embryonic stem cell research." Kerry has said he would not only remove the restrictions on funding of stem cell lines but also allow funding for therapeutic cloning, and that he would authorize an increase in its NIH funding from the current $24.8 million to $100 million. Asked whether Nancy Reagan had had reservations about her son's appearing at the Democratic Convention, Reagan said she had not. "She was aware that there would be dust rising — but she recognized the good opportunity it was for the issue she's so passionate about."

About ten days after Ron Reagan made his plea at the convention for a loosening of the restrictions on stem cell research — in what was the second-most-watched televised appearance, after Kerry's — Laura Bush, on a rare two-day campaign trip, issued a rebuttal, on the third anniversary of Bush's decision to limit federal funding. "Although you might not know about it from listening to the news lately, the president also looks forward to medical breakthroughs that may arise from stem cell research," she said, during a speech to the Pennsylvania Medical Society. "Few people know that George W. Bush is the only president to ever authorize federal funding for embryonic stem cell research." Laura Bush's statement was "technically true," said Jim Battey, of the NIH. "But it suggests that there were no plans to do it before him, and that's not true." The Clinton administration had, of course, gone to great lengths to make far fuller NIH funding possible — but Bush had prevented it. The distortion was, he added, an example of the kind of politicization that has plagued the field — "not necessarily lying but just selecting the facts that serve your own political purpose. And science is not generally well served by that kind of activity." Laura Bush's canard has been repeated by White House aides, in response to Kerry's attacks.

The heightened visibility of the issue in the national race might have a polarizing effect not only on Republican voters in California but also on its Republican governor. Before he ran for governor, Arnold Schwarzenegger and his wife, Maria Shriver, went to a dinner at the Zuckers' home, where an enthusiastic Maria plied the scientists with questions and Arnold expressed his support for embryonic stem cell research. After Ronald Reagan died, Schwarzenegger said again that he supported the research, but he has not taken a position on the initiative. The California Republican Party has opposed it. Moreover, it was only about nine months ago that Schwarzenegger, campaigning for another bond initiative — one that won on the March ballot, enabling the state to borrow $15 billion — swore that it was to pay down old debt, and after that there would be "no new debt." "Tear up your credit cards and throw them away!" he had said, ripping up an oversized card. How would he explain his support, on the very next ballot, for assuming another $6 billion in debt? Prop 71's advocates were told that Schwarzenegger probably could not support the measure and that he would stay neutral; but by late August he was reportedly under pressure from Republicans, both within and outside the state, to oppose it. About a week later, that pressure was said to have been "blunted," possibly by Maria. When I asked Klein about it, he grinned and said, "No comment." He also said that he made "a signal attempt to meet with Maria." In any event, Schwarzenegger's opposition is probably the Prop 71 proponents' greatest fear.

By late September, just as Prop 71's intensive television campaign began, a *Los Angeles Times* poll found that voters supported the measure 54 percent to 32 percent; 14 percent said they remained undecided. There was still a wild disparity in funding between the measure's supporters and its opponents. Representatives of the Catholic Conference and the Evangelical Christian Howard Ahmanson, Jr., began organizing the opposition to Prop 71 months ago, but they emerged publicly in their substantial roles only in September, when, records show, the Roman Catholic Church and Ahmanson were the two largest contributors to "No on

Prop 71," donating $50,000 each. The opposition has less than $200,000 in total; the Prop 71 campaign has $20 million. Klein says that he worries about a late outpouring of funds from the Christian right in the campaign's final days; he has put his house on the market and will sell it if the campaign needs last-minute funding.

It is ironic that John Kerry's impassioned advocacy of embryonic stem cell research could become an obstacle for its greatest boosters in California — but that's what happens when an issue has to win an election. And, at another level, some of Prop 71's backers might even have mixed emotions about a victory by the man who has sworn to remove the shackles from the NIH. California research would still have its pot of gold, but its chances of a near monopoly on that research would be substantially diminished.

When the kickoff event for the Prop 71 campaign was held at Cedars-Sinai Hospital in Los Angeles, on September 21, the ceremony reflected the organizers' belief that the argument most likely to persuade those who believe in the inviolability of the human embryo is the sight of suffering children and doomed adults. Volunteers held large blue posters before the TV cameras: YES ON 71 — SAVE LIVES WITH STEM CELLS. Bob Klein spoke. "My fourteen-year-old son, Jordan, has juvenile diabetes, and my mother is dying of Alzheimer's," he said. Klein has probably uttered these words hundreds of times in the past eighteen months; each time, he pauses briefly to compose himself. He then recited the growing list of supporters: twenty-two Nobel laureates, more than seventy medical and patient-advocacy organizations, and dozens of business organizations.

A soft-featured, serious-looking girl of thirteen named Emma Klatman stepped to the microphone. "I was diagnosed with type 1 diabetes when I was seven years old," she said. "Suddenly, I was no longer a naive young little girl, but a seven-year-old who had to test blood sugars with needles and give myself insulin shots for every morsel of food I ate. This is what I do every day and will have to for the rest of my life, if we do not find a cure. Vote for the cure!" Mark Siegel, a man in his fifties who spoke with some effort, said he had been diagnosed with Parkinson's disease six years ago. "I look forward to a future of shuffling down the street, not being able to smile the way I used to — eventually, losing control of my limbs." As for a cure, "they say it will take time, maybe five to ten years." He paused, and then said, more strongly and clearly, "I say, let's start the clock running now."

Dustin Hoffman, who had been standing on the small stage with other speakers, told the audience, "It's hard to believe, after listening to Mark and Emma, that we're even debating this. If each of them could be sent to every home in California, it might be their first unanimous vote. Listening to them, the question that appears in one's mind is, do we have to be afflicted to be enlightened?"

Those who believe in the personhood of the embryo and those who don't but who oppose creating an embryo for research do not, of course, subscribe to Hoffman's view of enlightenment. The real test of their convictions will come if the hoped-for cures materialize and if it is their loved ones who are afflicted but could be saved. That will be the crucible.

FREDERICK CREWS

Out, Damned Blot!

FROM *The New York Review of Books*

1

"It's a Rorschach." That bit of everyday speech, referring to any equivocal stimulus that elicits self-betraying interpretations on all sides, is one sign among many that, in the popular mind at least, the vaunted inkblot challenge has no rival as psychology's master test. In actuality, the Rorschach is now administered for diagnostic purposes somewhat less frequently than the low-maintenance, question-and-answer Minnesota Multiphasic Personality Inventory (MMPI), which asks the subject to agree or disagree with such flat-footed assertions as "I often feel sad." But neither the public nor Rorschachers, as the zealous and clannish guardians of the blot technique are known, take much interest in "superficial" self-report tests such as the MMPI. The mind's hidden layers, it is assumed, can be tapped only through unguided responses to images lacking determinate content; and the Swiss psychiatrist Hermann Rorschach's ten cards with bisymmetrical shapes, introduced to an initially unimpressed world in 1921, are thought to have confirmed their uncanny power in countless applications.

This judgment is shared in large measure by American clinical psychologists and other professionals who have occasion to administer personality tests. As we learn from a provocative and important book by James M. Wood, M. Teresa Nezworski, Scott O. Lilienfeld, and Howard N. Garb, *What's Wrong with the Rorschach?*, some 80 percent of American Ph.D. programs in clinical psychology still emphasize the Rorschach in required courses; 68 percent of specialist programs in educational psychology teach Rorschach technique; and the test is employed by roughly a third of all psychologists evaluating parents in custody cases, criminals facing sentencing or parole, and children who may or may not have been abused. Until very recently, testimony by Rorschach experts has gone largely unchallenged in our courts.

Necessarily, then, *What's Wrong with the Rorschach?* is not just a history of the test's evolution and periodic vicissitudes. It is also, and with increasing social concern as the story approaches the present, a continual assessment of the merits and pitfalls of projective testing. Since the four authors have themselves been participants in recent debate about the Rorschach, there is no pretense of neutrality here. But Wood and his colleagues do aim at objectivity and fairness, and if they err at all it is on the side of mercy. Readers of *What's Wrong* will find no more lucid primer on the requirements of scientific prudence as they relate to the authentication of psychological tests.

An avid reader of both Freud and his Zurich colleague Jung, Rorschach conceived of his test as a nonsectarian aid to psychoanalysis, impersonally determining an individual's "experience type" *(Erlebnistypus)* without presuming to favor one psychodynamic faction or another. The idea was to present all test takers with ten loose printed cards, half of them in black and white and half including some colors, displaying an identical sequence of images. For each card the test giver would ask in the most neutral tone, "What might this be?"; he would capture the subject's responses fully and exactly in a "protocol," or written record; and he would subsequently arrive at a singular personality profile by sifting that record for telltale features such as the kinds of forms named, whether the focus was on whole shapes or details, emphasis on movement versus color, and whether there was a rigid literalism or a comfortably imaginative accommodation to the imperfect resemblance between the blots and any real-life form.

The extent to which responses emphasized movement and color was of paramount importance in Rorschach's system of weighing personality. He believed that test takers who offer a high number of movement ("M") responses are, paradoxically, turned inward or "introversive"; intelligent and creative, they nonetheless are awkward and socially inept. In contrast, subjects who favor color ("C") responses are "extratensive," or adroit in company but restless and impulsive. Someone who registers a high number of both M and C scores qualifies as "dilated" or "ambiequal" — a healthy blending of introversive and extratensive traits. But low and similar numbers of M and C responses stigmatize the subject as "coarctative," or lacking in both creativity and emotional stability.

These rules were far from modest in scope. The close association between creativity and social clumsiness, were it to be upheld by evidence from other sources, would in itself constitute a major discovery, and so would the posited link between social adeptness and impulsiveness. In addition, it would be remarkable if those and other constellations could be inferred with certainty from such utterances as "The bug is bleeding" and "It looks like a skull." And this is to say nothing of Rorschach's most expansive boast, which was that his test would be found capable of ascertaining personality differences between regional populations and even whole races. Did he have grounds for making such sweeping claims, or was he capriciously assigning the equivalent of fortune cookies to his unsuspecting volunteers?

There is much in Rorschach's only book, *Psychodiagnostics,* that might encourage us to regard him as a crank. Bizarrely, for example, he insisted that a movement response be scored if the subject conjured a child sitting at a desk or a vampire sleeping in a coffin, because "muscular tension" was supposedly implied. And although a dog performing in a circus exhibited Rorschach movement, a cat catching a mouse or a fish darting through water did not, because, according to the founder, significant motion had to be "human-like" in function. Meanwhile, Rorschach tagged as "pedants" or "grumblers" any test takers who concentrated on details as opposed to whole images; those who interpreted white spaces were probably troublemakers; and those who hesitated before commenting on the multicolored cards must be exhibiting "color shock," thereby betraying themselves as neurotic repressers of emotion.

Rorschach argued, quite sensibly, that by examining the average test results (later called "norms") for many people whose personality traits had already been determined by other means, administrators could learn whether a given kind of response was actually well correlated with a given trait. Yet before he died from a perforated appendix just nine months after the publication of *Psychodiagnostics,* Rorschach had found time to accumulate test results for only 405 independently categorized subjects, and their types were drastically skewed toward schizophrenia (188 examples) and other pathologies he had encountered in his hospital rounds. Only 117 ordinary people, scattered amid his assorted "morons," "imbeciles," "senile dements," and so forth, had been sampled. Given such sketchy evidential base, it isn't surprising that *Psychodiagnostics* was slow to find admirers; the wonder is that its complex scoring system was finally adopted with so few reservations.

The Rorschach found its true welcome in the world's headquarters of psychological typecasting and "adjustment," the United States. Wood engagingly tells how the test finally caught on here in the 1930s, flourished in the forties and early fifties, weathered a crisis of doubt in the later fifties and sixties, and then surged again until, beginning a decade ago, skeptics began to nip at its heels once more. Along the way, different groups of American enthusiasts devised their own scoring rules to yield the kinds of results that interested them. Through it all, however, Rorschachers have kept faith with the founder's ten inviolate cards, which have been granted the kind of awe once reserved for texts dictated directly from the sky.

The Rorschach conquered North America and much of the Western world before any part of its rationale had been subjected to stern experimental trial. In seeking to explain this striking fact, Wood notes that inkblot games and tests were current even before Rorschach launched his own version. His key departure — the attempt to gauge a subject's whole personality and not just a faculty of imagination — fit nicely with the growing sway of psychoanalysis, and more particularly with the Freudian ideas of projection and free association. Again, Americans who preferred the more cheerful Jungian conception of the psyche responded favorably to Rorschach's adaptation (with significant differences) of Jung's already celebrated dichotomy between introverts and extroverts. Moreover, in balancing a "romantic" emphasis on deep intuition against an "empiricist" battery of codes and tables for figuring scores, the Rorschach proved at first serviceable, and then virtually indispensable, to the burgeoning American profession of clinical psychology, which was developing its own romantic pretensions but needed an objective-looking diagnostic tool to offset the inherent subjectivism of the one-on-one interview.

2

Reasons for popularity, of course, are not the same thing as scientific justifications. Wood and his coauthors remind us that if a given instrument of testing in any field is not to cause havoc, it must be both valid and reliable. In brief, it must measure what it purports to measure and it must yield approximately the same results when readministered in new conditions or by other examiners.

Hermann Rorschach had accepted those criteria in principle, and most of his followers have paid due obeisance to them. But at every juncture where the test stood in peril of being decertified by negative findings, Wood shows, its promoters backed off from empirical accountability and expanded the scope of their claims. The story told in *What's Wrong*, by turns appalling and amusing, reads like a parable of the larger struggle between science and pseudoscience, with the latter always managing somehow to issue itself a new reprieve from execution.

When the Rorschach began to attract American followers through word of mouth in the 1930s, it brought to prominence an initially reluctant but subsequently flamboyant champion, Bruno Klopfer, whose talent for salesmanship and deafness to criticism were responsible in part for the high morale of American Rorschachers in the forties and fifties. A refugee from Nazi Berlin, Klopfer had studied with Jung in Zurich and had learned how to score various psychological tests there, including the Rorschach. He was barely surviving as a research assistant in Columbia's anthropology department when eager graduate students learned of his expertise and pressed him into moonlighting as their Rorschach trainer.

Although Klopfer's real passion had been psychoanalysis, not assessment, he soon contracted a taste for interpreting Rorschach's still untranslated pronouncements and for devising novel inkblot rules that hadn't occurred to the master. Before long he possessed a grand career and an adoring crew of disciples who fed his insatiable ego. As Wood explains, this elevation of one person to guruhood added mystification to an already dubious mind-reading program and further postponed a reckoning with the need for evidential support.

In the Rorschach scheme as first conceived, the sum of a subject's scores for responses in each category of interest — color, say, or white-space shapes — corresponded directly to a certain trait of personality. Klopfer accepted some of those equivalences, but on the whole he found the idea behind them too rigid for capturing the subtleties of human character. What was needed, he argued, was "configural" interpretation, whereby a highly experienced and gifted judge (guess who?) would draw "holistic" inferences from an intuitive contemplation of all of the subject's scores on the test. The Rorschach judge or "artist" could justify this method by creating anonymous ("blind") profiles on the basis of protocols compiled by others and then by checking the profiles against case histories or against delayed personal acquaintance with the test takers. The artist himself or someone from his circle of admirers would let the rest of us know, anecdotally, how well he had done.

Klopfer's fans considered him a virtual oracle, and he was inclined to agree. He even claimed that, through analysis of Rorschach scores alone, he could discriminate between cancer patients with fast- and slow-growing tumors. But all of his pretensions were proven hollow when neither he nor other famous virtuosos could exhibit any diagnostic acumen in circumstances that were properly secured against cheating. Though many Rorschachers still revere Klopfer's memory, he now appears to have been only a colorful buffoon.

Wood and his fellow research psychologists are only peripherally concerned with Klopfer's foibles. Their target here is the whole idea of "clinical validation," whereby hypotheses are checked not against impersonal trials of their adequacy but against testimonials, case studies, and assessments of success made by parties with a stake in the outcome. That method risks being undermined by "confirmation bias," or the natural human tendency to misread evidence in one's favor. And confirmation bias runs wild in subjective evaluations of Rorschach profiles, thanks to such factors as the test's excessively broad classification of traits and the contradictions that crop up within a given subject's scores, tempting the evaluator to seize upon apparent hits and ignore the misses. Even blind

Rorschach interpretation per se, Wood points out, isn't always what it seems, because the examiner often has advance knowledge about the population of test takers — for example, a ward full of mental patients.

Bruno Klopfer's disdain for controlled studies was shared by American psychoanalysts, whose own absolute trust in clinical validation had come straight from Freud. The Rorschach, they perceived, could serve as a technical adjunct to their relatively unstructured explorations of patients' minds. Predictably, they imbued the test with a symbol-decoding function — an approach that Rorschach himself, decades earlier, had pondered and rejected as unproductive. According to the Freudians, subjects gazing at the ten plates were really seeing projections of their unconscious desires and neuroses — conditions that pointed to a need for further months or years on the couch. And needless to say, those inferences, too, were clinically validated without incurring any risk of disconfirmation.

At the height of the Freudian vogue, Rorschach's Cards IV and VII became known to analytically inclined authorities, though not to unsuspecting test takers, as the "Father" and "Mother" images. A woman who pointed out, plausibly, that the "arms" on the Father card are skinny was said to be in the grip of penis envy; and if she likened the Mother to a stuffed animal, she thereby convicted herself of what one expert called "a refusal to grow up and assume heterosexual responsibilities." "So it went with all the cards," Wood observes. "Card V revealed childhood memories of having seen one's parents engaged in intercourse. Card VI reflected unconscious attitudes toward sex and 'phallic worship.' Card IX revealed 'anal' concerns and paranoia. Card X revealed 'oral' fantasies."

In this vein of instant diagnosis no one surpassed the psychoanalyst Robert Lindner, the author of *Rebel Without a Cause*. Lindner identified forty-three Rorschach responses amounting to cries for help. Does the subject perceive "some sort of tool" in Card II? Then he is suffering from "hesitancy in coming to grips with an underlying sexual problem." Does he compare a portion of Card X to an extracted tooth? He is a chronic masturbator. And if, when perusing Card IV, he is rash enough to mention both decay and death, he is probably suicidal, and "there is a fair prospect that [he] will benefit from convulsive therapy."

All of the eminent American Rorschachers from the thirties

through the fifties were sympathetic to Freudian dream interpretation, and they felt the tug of the popular psychoanalytic tide. After
considerable hesitation, for example, and not without misgivings
about diluting his authority, Bruno Klopfer made room in his scoring system for Mother and Father symbolism. Yet the main benefit
of the Freudian trend accrued to those tradition-minded Rorschachers who resisted it. In doing so, they identified themselves as
the party of scientific restraint — even though their own rules for
interpreting color, movement, and form responses as indicators of
personality had never been proven cogent, either.

From the outset of the Rorschach's wild American ride, some psychologists who believed in the test's general soundness understood
that empirical standards couldn't be indefinitely brushed aside in
the lordly Klopfer manner. Two highly regarded Rorschach theo-
rists, Samuel Beck and Marguerite Hertz, argued vigorously that
Klopfer's "configural" blending of scores was perpetuating a deadly
subjectivism and placing each individual rule of interpretation beyond the reach of disproof. Beck and Hertz won an appreciative
following by associating themselves with "psychometrics," the statistically based controls that are now universally honored in experimental psychology if not in its clinical counterpart. As they emphasized, the psychometric ethos mandates that test procedures be
standardized; that reliability be verified, not just promised; and
that norms be gathered in enough volume to put the announced
meaning of scores beyond dispute.

Beck and Hertz could endorse psychometrics because they were
initially sure that research would confirm most Rorschach rules
while weeding out a few unsupportable ones. That confidence inspired many loyalists as well as outsiders, from the forties through
the sixties, to submit Rorschach hypotheses to objective review.
But the results proved devastatingly negative. Correlations between
predicted traits and independently observed ones were found to
be either nonexistent or too weak to be trusted. Test results varied
unacceptably among examiners with differing styles of self-pres-
entation. And meanwhile, crippling statistical blunders were un-
earthed in the more optimistic reports. As the most sophisticated
and persistent of the critics, Lee J. Cronbach, wrote in 1956, "It is
not demonstrated that the test is precise enough or invariant

enough for clinical decisions. The test has repeatedly failed as a predictor of practical criteria . . . There is nothing in the literature to encourage reliance on Rorschach interpretations."

Here was one of those moments when Rorschachers faced a clear if painful choice between loyalty to science and loyalty to beliefs in which they had invested much money, time, and self-esteem. As usual, most of them opted to close ranks. Even Samuel Beck ignored the ominous findings and began taking a kinder view of clinical validation after all. And most of those who did refer to the experimental literature, having sifted it for scraps of encouragement, refused to acknowledge what it was plainly saying about the test's fundamental inaccuracy. When practitioners of a quasi-medical fad exempt themselves from answerability to empirical trials, one usual consequence is a proliferation of alternative schools. By the late 1950s, Wood reports, no fewer than five American Rorschach regimens were current, not to mention other inkblot tests that heretically departed from the original cards. Credentialed psychologists, rapidly increasing in number but not in methodological sophistication, felt free to draw upon all five incompatible codes as if they were passing down the line at a salad bar. To the increasingly restive dissenters, what had been true all along was now overwhelmingly apparent: the Rorschach was a revealing projective test not of its respondents' quirks but of the preconceptions held by its advocates.

3

By the mid-1960s the Rorschach had earned the distrust of most psychologists who were keeping up with the mainstream journals. In 1974, however, the technique underwent a surprising resuscitation at the hands of the American clinician John E. Exner, whose often revised and supplemented book, *The Rorschach: A Comprehensive System,* would become the most influential of all Rorschach texts.

Seeking consensus among the quarreling Rorschachers, Exner assembled an eclectic quilt of the best elements, as he judged from painstaking surveys, in all of the competing regimens while adding many categories of his own, including new measures for egocentricity, depression, obsessive style, and "hypervigilance." In itself, such a combination of winnowing and amplifying wouldn't have set Exner apart from many another Rorschach pundit. But he silenced most doubters by conspicuously embracing psychometric standards, including the provision of abundant, broadly representative norms and barrages of scientific references purporting to document both the validity and the reliability of his Comprehensive System.

In Exner's hands the test regained a level of respect not enjoyed since the early fifties, and his domination of the Rorschach scene has been all but total for three decades now. In 1997 he received an Award for Distinguished Professional Contributions to Knowledge from the Board of Professional Affairs of the American Psychological Association, which credited his Comprehensive System with having revived "perhaps the most powerful psychometric instrument ever envisioned"; and as recently as April 2004 his life's work was honored through a conference jointly sponsored by Harvard Medical School and the Massachusetts Mental Health Center.

As *What's Wrong with the Rorschach?* demonstrates, however, the imposing Comprehensive System is really a production of smoke and mirrors. Exner's claims for the high reliability of his technique, it turns out, have rested on a misconstrual of accepted statistical terminology. Further, his famous compilation of norms was vitiated by a major sampling error that went unremarked for more than a decade. The cited studies underpinning his rules have been mostly unpublished and unshared work by the least trustworthy of judges, a team of enthusiasts employed by his own subsidiary, Rorschach Workshops, Inc. And the inflated reputation of those studies has been sustained by the Rorschach Research Council, yet another Exner satellite.

In Exner's code, "reflection" answers, such as "clouds reflected in a pond," indicate narcissism. Even one reflection response within a protocol, Exner admonishes, can tell a psychologist that "a nuclear element in the subject's self-image is a narcissistic-like feature that includes a marked tendency to overvalue personal worth." This is a grave matter, because Exner, taking a page from Freud, has held that "homosexuals and sociopaths" are highly narcissistic. Detection of those dubiously bracketed personality types thus rests in part on the perception of symmetries in forms that simply *are* symmetrical.

Wood's critique leaves me convinced that Exner's pseudo-precise method for inferring diagnoses from weighted combinations of Rorschach scores is even riskier than Bruno Klopfer's impressionistic holism. Take, for instance, the Comprehensive System's Egocentricity Index, which is figured by tripling the number of a subject's reflection responses, adding the number of "pair" responses, and dividing the sum by the total number of all responses. If any one of the component rules here is invalid, so is the whole index, in which case its automatically applied mathematical formula will cause many more misdiagnoses than Klopfer's case-by-case guesswork.

The Comprehensive System has done less than nothing to remove the most serious flaw in all previous Rorschach schemes, including Hermann Rorschach's own: their tendency to overpathologize, or to err on the side of abnormal characterizations. In one study from the 1980s, in which mental patients and ordinary citizens were blindly intermingled, Comprehensive System judges classified nearly 80 percent of the normal subjects as depressed and as harboring serious problems of character. And in 2000, three skeptical researchers, examining the Rorschach scores of one hundred behaviorally normal California schoolchildren, reported that, according to the Comprehensive System, the children

> may be described as grossly misperceiving and misinterpreting their surroundings and having unconventional ideation and significant cognitive impairment. Their distortion of reality and faulty reasoning approach psychosis. These children would also likely be described as having significant problems establishing and maintaining interpersonal relationships and coping within a social context. They apparently suffer from an affective disorder that includes many of the markers found in clinical depression.

When the Exner system is relied upon to assess individual children or candidates for the priesthood or pilots suspended for drunkenness or convicts seeking parole, a form of roulette is being played with their fate. James Wood offers a chilling example from a custody dispute in which he himself was consulted too late to affect the outcome. Inkblot scores had suggested that the ex-wife, who had repeatedly charged her ex-husband with physically and sexually abusing his children, was seriously disturbed, lacking in empathy, and incapable of forming rational judgments. Those were the conclusions implied by such damning symptoms as her having seen, in two cards, the shape of a paper snowflake (incorrectly scored by the examiner as a "reflection response") and the supposedly depressive image of a Thanksgiving turkey carcass — corresponding, as it happened, to leftovers that were then sitting in her refrigerator.

Meanwhile, the ex-husband's Comprehensive System protocol assured the authorities that he was more or less normal. As Wood learned from available records, however, the man had beaten at least one of his three previous wives, had married this fourth one under an assumed name, had broken two of her teeth shortly after their wedding, and had both battered and molested his young son. Thanks in large part to the two errant Rorschach profiles, full custody of the endangered son was awarded to his sadistic father.

If a psychological test cannot discriminate reliably between signs of pathology and casual associations such as the remembered turkey carcass, it is a public menace and ought to be dropped forthwith. Oddly, however, Wood and his coauthors are reluctant to say so. Partly out of deference to colleagues who have devoted their careers to the Rorschach and partly because some of the authors themselves still harbor confessed "romantic" sympathies, they take a firm stand only about the urgency of getting the Rorschach out of the courtroom. On other points, such as the possible usefulness of inkblots in "psychodynamic exploration" that could be "analogous to dream interpretation," they express cautious interest and call for further research.

Here the authors appear to have put in abeyance their own decisive critique of clinical validation. Both dreams and Rorschach responses can be "explored" with disastrous effect; think of the role played by dream analysis in recovered memory therapy, and think of Robert Lindner's suggestion that shock treatment may be indicated when Rorschach answers reveal a desperate mental state. The story told with admirable patience and logic in *What's Wrong with the Rorschach?* speaks more clearly than its authors do here at the end. This test is a ludicrous but still dangerous relic of the previous century's histrionic love affair with "depth," and the only useful purpose it can serve now is as a caution against related follies.

JARED DIAMOND

Twilight at Easter

FROM *The New York Review of Books*

1

NO OTHER SITE that I have visited made such a ghostly impression on me as did Rano Raraku, the quarry on Easter Island where its famous gigantic stone statues were carved. To begin with, the island is the world's most remote habitable scrap of land, lying far out in the southeastern Pacific Ocean, 2,300 miles west of the coast of Chile and 1,300 miles east of Pitcairn Island. Rano Raraku itself is a volcanic crater 600 yards in diameter, which I entered by a trail rising steeply up to the crater rim from the plain outside, and then dropping steeply down again toward the marshy lake on the crater floor. No one lives in the vicinity today.

Scattered over the crater's walls are 397 stone statues, each representing in a stylized way a long-eared legless human male torso, most of them 15 to 20 feet tall, the largest of them 70 feet tall (taller than the average modern five-story building), and weighing from 10 to 270 tons. The remains of a transport road can be discerned passing out of the crater through a notch cut into a low point in its rim, from which three more roads radiate north, south, and west for up to nine miles toward Easter's coasts.

Scattered along the roads are 97 more statues, as if abandoned in transport from the quarry. Along the coast are 113 stone platforms that formerly supported or were associated with 393 more statues, all of which (until the recent re-erection of a few) were no longer standing but had been thrown down, many of them toppled and deliberately broken at the neck. Yet Easter Island's Polynesian population had possessed no cranes, wheels, machines, metal tools, draft animals, or means other than human muscle power to move the statues.

Statues remaining at the quarry are in all stages of completion. Some are still attached to the bedrock out of which they were carved, roughed out but with details of the ears or hands missing. Others are finished, detached, and lying on the crater slopes below, and still others had been erected in the crater. Littering the ground everywhere at the quarry are the stone picks, drills, and hammers with which the statues were being carved. The scene gave me the sense of a factory all of whose workers had suddenly quit for mysterious reasons, thrown down their tools, and stomped out, leaving each statue in whatever stage it happened to be at the moment. Who carved the statues, how did the carvers move such huge stone masses, and why did they eventually throw them all down?

Easter's mysteries were already apparent to its European discoverer, the Dutch explorer Jacob Roggeveen, who spotted the island on Easter Day (April 5, 1722), hence the name that he bestowed and that has stuck. Like all subsequent visitors, Roggeveen was puzzled, not understanding how the islanders had transported and erected their statues. No matter what had been their exact method, they would have needed heavy timber and strong ropes made from big trees, as Roggeveen realized. Yet the Easter Island that he viewed was a wasteland with not a single tree or bush over ten feet tall. What had happened to all the trees that must have stood there?

All those mysteries have spawned volumes of speculation for almost three centuries. Many Europeans were incredulous that Polynesians, "mere savages," could have constructed the statues or the beautiful stone platforms. The Norwegian explorer Thor Heyerdahl's famous *Kon-Tiki* expedition and his other raft voyages aimed to prove the feasibility of transoceanic connections among Egypt's pyramids, the giant stone architecture of South America's Inca Empire, and Easter Island's statues. Going further, the Swiss writer Erich von Däniken claimed that the statues were the work of intelligent extraterrestrials who had ultramodern tools, became stranded on Easter, and were finally rescued. But the explanation that has now emerged attributes statue carving to the picks and other tools littering Rano Raraku rather than to hypothetical space implements, and to Easter's known Polynesian inhabitants rather than to Incas, Egyptians, or Martians. This story is as romantic and exciting as were postulated visits by *Kon-Tiki* rafts or extraterrestrials — and much more relevant to events now going on in the modern world.

Easter's history has recently been recounted in two excellent but very different books, both by the authors of the two previous standard books about the island. The geographer and botanist John Flenley, who uncovered the evidence for Easter Island's vanished forest and extinct giant palm trees, has collaborated with the well-known archaeologist Paul Bahn to bring the Easter Island story up to date with the discoveries of the last decade. They thereby offer us clear summaries of Easter's settlement and subsequent history, its statues, the frightening collapse of its society, and its broader significance in our world beset with similar environmental problems.

The archaeologist Jo Anne Van Tilburg, the leading authority on the statues themselves, has used her understanding of Easter Island history and statues in her biography of the remarkable self-trained archaeologist and ethnographer Katherine Routledge, who spent seventeen months on the island in 1914–1915, and whose unpublished handwritten field notes Van Tilburg deciphered. The information in those notes is of lasting value because Routledge was an excellent interviewer, and some of her older informants had participated in the island's last traditional ceremonies (the so-called Orongo birdman rites). Those informants told Routledge masses of information about traditional Easter Island society that would otherwise be lost to us.

Van Tilburg has really given us three books in one: a history of a unique society, a Gothic novel, and a powerfully moving biography. The variously furious, passive-aggressive, inept, and effective relations of Routledge and her husband with each other, with other expedition members, with islanders, and with the island priestess Angata, who gained spiritual power over Routledge — all that makes a fascinating story. Routledge wrote of herself in 1891, "It was my misfortune to be born a woman with the feelings of a man." Her tragic biography traces how a rich heiress with a family history of mental illness mastered her inner problems sufficiently to become one of the earliest women graduates of Oxford University, then to make her own way through a man's world, and to contribute to our understanding of Easter Island, only to succumb at last to paranoia and to die in the mental asylum to which her husband and brother finally committed her.

2

From Flenley and Bahn's and Van Tilburg's accounts, it becomes clear how both Heyerdahl and von Däniken brushed aside overwhelming evidence that the Easter Islanders were indeed typical Polynesians, speaking a Polynesian language and making stone tools in the usual Polynesian styles. Around AD 900 they colonized Easter Island from Polynesian islands to the west and built up a population that peaked at around 15,000 people. At the time of the European arrival they were subsisting mainly as farmers, growing yams, taro, bananas, sugar cane, and sweet potatoes, as well as raising chickens, their sole domestic animal. While Easter Island was divided into about eleven territories, each belonging to one clan under its own chief and competing with other clans, the island was also loosely integrated religiously, economically, and politically under the leadership of one paramount chief. On other Polynesian islands, competition between chiefs for prestige could take the form of inter-island efforts such as trading and raiding, but Easter's extreme isolation from other islands precluded that possibility. Instead, the excellent quality of Rano Raraku volcanic stone for carving eventually resulted in chiefs competing by erecting statues representing their high-ranking ancestors on rectangular stone platforms (termed *ahu*).

Each of the island's eleven territories contained between one and five large *ahu* up to 13 feet high, many extended by lateral wings to a width of up to 500 feet. Today the *ahu* are a dingy dark gray, but originally they must have been a colorful white, yellow, and red: the facing slabs were encrusted with white coral, the stone of a freshly cut statue was yellow, and the statue's crown and a horizontal band of stone coursing on the front wall of some *ahu* were red.

The *ahu*-building period seems to have begun around AD 1000 or 1100, within a few centuries of the island's settlement. An increase in statue size with time suggests competition between rival chiefs commissioning statues to outdo each other. (In case that strikes you as weird, try imagining what a dispassionate observer would say about the increasingly lavish cars, mansions, and jewelry by which modern American "chiefs" compete.) The strong possibility of such competition also seems evident from an apparently late feature called a *pukao:* a cylinder of red volcanic stone, weighing up to twelve tons, mounted as a separate piece to rest on top of a statue's flat head, and possibly representing a chief's headdress or hat of red feathers. All *pukao* are from a single quarry, Puna Pao, where (just as with the statues themselves in Rano Raraku quarry) I saw unfinished *pukao*, plus finished ones awaiting transport. We know of only about sixty *pukao*, reserved for statues on the biggest and richest *ahu*. I cannot resist the thought that they were produced as a show of one-upmanship. They seem to proclaim: "All right, so you can erect a statue 32 feet high, but look at me: I can lift this 12-ton pukao on top of my statue; you try to top that, you wimp!"

How did the islanders succeed in erecting and transporting those statues? Of course we don't know for sure, because no European ever saw it being done to write about it. But we can make informed guesses from the oral traditions of the islanders themselves and from recent experimental tests of different transport methods described by Flenley and Bahn and carried out and described by Van Tilburg.

The still-visible transport roads on which statues were moved from Rano Raraku quarry follow contour lines to avoid the extra work of carrying statues up and down hills and are up to nine miles long for the *ahu* farthest from the quarry. While the task may strike us as daunting, we know that many other prehistoric peoples transported very heavy stones at Stonehenge, Egypt's pyramids, and Inca and Olmec centers, and something can be deduced of the methods in each case. The method most convincing to me is Van Tilburg's suggestion that Easter Islanders modified the so-called canoe ladders widespread on Pacific islands for transporting heavy wooden logs, which had to be cut in the forest, shaped into canoes, and then transported to the coast.

The "ladders," which I have seen on islands near New Guinea, consist of a pair of parallel wooden rails joined by fixed wooden crosspieces over which the log is dragged. We know that some of the biggest canoes that the Hawaiians moved over such horizontal ladders weighed more than an average-sized Easter Island statue, so the proposed method is plausible. Van Tilburg persuaded modern Easter Islanders to put her theory to a test by building such a canoe ladder, mounting a statue prone on a wooden sled, attaching ropes to the sled, and hauling it over the ladder. She found that between fifty and seventy people, working five hours per day and dragging the sled five yards at each pull, could transport an average-sized twelve-ton statue nine miles in a week. By extrapolation, transport of even the biggest statues could have been accomplished by a team of five hundred adults, which would have been just within the manpower capacities of an Easter Island clan.

Islanders told Thor Heyerdahl how their ancestors had erected statues on an *ahu;* they were indignant that archaeologists had never deigned to ask them, and to prove their point they erected a statue for him without a crane. They began by building a gently sloping ramp of stones up to the top of the platform and pulling the prone statue with its base end forward up the ramp. Once the base had reached the platform, they levered the statue's head an inch or two upward with logs, slipped stones under the head to support it in the new position, and continued to lever up the head and thereby tilt the statue increasingly toward the vertical.

However, we have glossed over a problem. Transporting and erecting statues required lots of thick long ropes (made in Polynesia from fibrous tree bark) to drag the sleds and heavy statues, and also many big strong trees to obtain all the timber needed for the sleds, canoe ladders, and levers. But the Easter Island seen by Roggeveen and subsequent European visitors had very few trees, all of them slight and short: it is the most nearly treeless island in Polynesia. Where were the trees that provided the required rope and timber?

3

Botanical surveys of plants living on Easter Island within the twentieth century have identified only forty-eight native species, even the biggest of them hardly worthy of being called a tree (just seven feet tall), and the rest of them low ferns, grasses, sedges, and shrubs. However, beginning especially with John Flenley's and Sarah King's studies in 1984, several methods for recovering and identifying pollen and wood charcoal from vanished plants have shown that, long before human arrival and still during the early days of human settlement, Easter was not a barren wasteland but supported a subtropical tall forest.

As Flenley and his colleagues recognized, the most interesting of those extinct trees was what used to be the world's largest palm tree, related to but dwarfing the largest existing palm, the Chilean wine palm, which grows up to 65 feet tall and 3 feet in diameter. Chileans prize their palm today for several reasons, and Easter Islanders would have done so as well. As the name implies, the trunk yields a sweet sap that can be fermented to make wine or boiled down to make honey or sugar. The nuts' oily kernels are a delicacy. The fronds are ideal for fabricating into house thatching, baskets, mats, and boat sails. And of course the stout trunks would have served to transport and erect statues and to make rafts.

Many of the twenty-one other vanished plant species besides the palm would also have been valuable to the Easter Islanders. Two of them are tall trees used elsewhere in Polynesia for making canoes. The bark of one of them is used by Polynesians to make rope, and that was presumably how Easter Islanders dragged their statues. Still others variously yielded bark cloth, edible fruits, firewood, or hard wood good for carving, construction, and making harpoons.

Studies of vertebrate bones from middens — mounds of shells, bones, and other refuse — at the probable site of the first human settlement prove that Easter, which today supports not a single species of native land bird, was formerly home to at least six of them, including one species of heron, two chickenlike rails, two parrots, and a barn owl. More impressive was Easter's prodigious total of at least twenty-five nesting sea bird species, making it formerly the richest breeding site in Polynesia and probably in the whole Pacific. They must have been attracted by Easter's remote location and lack of predators, which made it a safe haven as a breeding site — until humans arrived.

The excavations that yielded those bones tell us much about the diet and lifestyle of Easter's early human settlers. The most frequent bones, accounting for more than one-third of the total, be-

long to the largest animal available to Easter Islanders: the common dolphin, weighing up to 165 pounds. That's astonishing:
nowhere else in Polynesia do dolphins account for even as much as
1 percent of the bones in middens. The dolphin generally lives out
to sea, hence it could not have been hunted by line-fishing or
spearfishing from shore. Instead, it must have been harpooned far
offshore, in big seaworthy canoes built from the now-extinct tall
trees. Fish bones and shellfish occur in the middens but in only
modest quantities, because Easter's rugged coastline and the steep
drop-off of the ocean bottom provide few places to catch fish or
shellfish in shallow water. To compensate, there were those abundant sea birds plus the land birds.

Comparison of early garbage deposits with late prehistoric ones
or with conditions on modern Easter Island reveals big changes in
those initially bountiful food sources. Porpoises and open-ocean
fish like tuna virtually disappeared from the islanders' diet. The
fish that continued to be caught were mainly inshore species. Land
birds disappeared completely from the diet, for the simple reason
that every species became extinct from some combination of over-
hunting, deforestation, and predation by rats introduced accidentally as stowaways in the colonists' canoes. This was the worst catastrophe to befall Pacific island birds, surpassing even the record on
New Zealand and Hawaii, where, to be sure, the moas and most
flightless geese became extinct, but many other species managed
to survive. No Pacific island other than Easter ended up without
any native land birds. Of the twenty-five or more formerly breeding
sea bird populations, overharvesting and rat predation brought the
result that only one now breeds on Easter itself. Even shellfish were
overexploited, so shell sizes in the middens decreased with time because of preferential overharvesting of larger individuals.

The giant palm and all the other now extinct trees disappeared
for half a dozen reasons that we can document or infer. Identified
tree charcoal fragments from ovens prove directly that trees were
being burned for firewood. Trees were being cleared for gardens,
because most of Easter's land surface ended up being used to grow
crops. From the early midden abundance of bones of open-ocean
porpoises and tuna, we infer that big trees were being felled to
make seaworthy canoes; the frail, leaky little watercraft seen by
early European visitors would not have served for harpooning plat-

forms or for venturing far out to sea. Trees furnished the timber and rope not only for transporting and erecting statues, but undoubtedly for a multitude of other purposes. The introduced rats "used" the palm tree and doubtless other trees for their own purposes: every Easter palm nut that has been recovered shows tooth marks from rats gnawing on it and would have been incapable of germinating. From several types of archaeological evidence, we deduce that the clearing of forests began soon after human arrival, reached its peak around 1400, and was virtually complete by dates that varied locally between the early 1400s and the 1600s.

The overall picture for Easter is the most extreme example of forest destruction in the Pacific, and among the most extreme in the world: the whole forest gone, and all of its tree species extinct. Immediate consequences for the islanders were losses of raw materials, losses of wild-caught foods, and decreased crop yields.

Raw materials lost or else available only in greatly decreased amounts consisted of everything made from native plants and birds, including wood, rope, bark to manufacture bark cloth, and feathers. Lack of large timber and rope brought an end to the transport and erection of statues, stopped the construction of seagoing canoes, and left people without wood for fires to keep themselves warm during Easter's winter nights of wind and driving rain at a temperature of 50 degrees Fahrenheit. Instead, after 1650 the islanders were reduced to burning herbs, grasses, and crop wastes for fuel. There would have been fierce competition for the remaining woody shrubs among people trying to obtain thatching and small pieces of wood for houses, implements, and bark cloth.

Most sources of wild food were lost. Without seagoing canoes, the bones of porpoises, tuna, and pelagic fish vanished from middens by 1500. The numbers of fishhooks and fish bones in general also declined, leaving mainly just fish species that could be caught in shallow water or from the shore. Land birds and wild fruits vanished from the list, sea birds were reduced to relict populations, and the shellfish consumed became fewer and smaller. The only wild food source whose availability remained unchanged was rats.

In addition to those drastic decreases in wild food sources, crop yields also decreased, for several reasons. Deforestation led locally to soil erosion by rain and wind, as shown by huge increases in the quantities of soil-derived metal ions carried into Flenley's swamp sediment cores. Other damages to soil that resulted from deforestation and caused lower crop yields included desiccation, nutrient leaching, and reduced rainfall. Farmers found themselves without most of the wild plant leaves, fruit, and twigs that they had been using as compost.

Those were the immediate consequences of deforestation and other human environmental impacts. The further consequences were starvation, a population crash, and a descent into cannibalism. Surviving islanders' accounts of hunger are graphically confirmed by the proliferation of little statues called *moai kavakava*, depicting starving people with hollow cheeks and protruding ribs. Captain Cook in 1774 described the islanders as "small, lean, timid, and miserable." Numbers of house sites in the coastal lowlands, where almost everybody lived, declined drastically in the 1700s from peak values between approximately 1400 and 1600, suggesting a corresponding decline in numbers of people. In place of their former sources of wild meat, islanders turned to the largest hitherto unused source available to them: humans, whose bones became common not only in proper burials but also (cracked to extract the marrow) in late Easter Island garbage heaps. Oral traditions of the islanders are obsessed with cannibalism; the most inflammatory taunt that could be snarled at an enemy was "The flesh of your mother sticks between my teeth."

Easter Island's chiefs and priests had previously justified their elite status by claiming relationship to the gods and by promising to deliver prosperity and bountiful harvests. They buttressed that ideology with monumental architecture and ceremonies designed to impress the masses, and made possible by food surpluses extracted from the masses. As their promises were being proved increasingly hollow, the chiefs and priests were overthrown around 1680 by military leaders called *matatoa*, and Easter's former complexly integrated society collapsed in an epidemic of civil war. The obsidian spear-points from that era of fighting still littered Easter in modern times. For safety, many people turned to living in caves whose entrances were partly sealed to create a narrow tunnel for easier defense.

What had failed, in the twilight of Easter's Polynesian society, was not only the old political ideology but also the old religion, which became discarded along with the chiefs' power. Oral traditions record that the last *ahu* and statues were erected around 1620. Around 1680, at the time of the military coup, rival clans switched from erecting increasingly large statues to throwing down each other's statues by toppling them onto a slab placed so that the statue's neck would fall on the slab and break. The last observation of an erect statue was in 1838.

Ahu themselves were desecrated by pulling out some of the fine slabs in order to construct garden walls or burial chambers. As a result, today the *ahu* that have not been restored (that is, most of them) look like mere boulder heaps. When I drove around Easter, I saw *ahu* after *ahu* as a rubble pile with its broken statues. I reflected on the enormous effort that had been devoted for centuries to constructing them, and then remembered that it was the islanders themselves who had destroyed their own ancestors' work. I was filled with an overwhelming sense of tragedy. Easter Islanders' toppling of their ancestors' statues reminds me of Russians and Romanians toppling the statues of Stalin and Ceauşescu when the Communist governments of those countries collapsed. The islanders must have been filled with pent-up anger at their leaders for a long time, as we know that Russians and Romanians were.

4

Why were Easter Islanders so foolish as to cut down all their trees, when the consequences would have been so obvious to them? This is a key question that nags everyone who wonders about self-inflicted environmental damage. I have often asked myself, "What did the Easter Islander who cut down the last palm tree say while he was doing it?" Like modern loggers, did he shout "Jobs, not trees!"? Or "Technology will solve our problems, never fear, we'll find a substitute for wood"? Or "We need more research, your proposed ban on logging is premature"?

Similar questions arise for every society that has inadvertently damaged its environment, including ours today. It turns out that there is a series of reasons why people in any society — whether Easter Islanders, Maya, or ourselves — may make fatal mistakes that will look foolish to their successors. They may not anticipate a problem because the problem is unprecedented in their experience: for example, today's overharvesting of the ocean's seemingly inexhaustible fisheries, for the first time in human history. They may fail to perceive the problem when it does arrive: for example, global warming today, initially difficult to distinguish from just the usual year-to-year fluctuations in temperature. Conflicts of interest may prevent them from addressing a perceived problem: for example, dumping toxic wastes into rivers is bad for people living downstream but saves money for the company doing the dumping. Some problems just prove too difficult to solve with current abilities: for example, no one has figured out how to eliminate the Dutch elm disease that reached North America. Probably all of those kinds of explanations apply to deforestation on Easter Island, but the most important reason there may be conflicts of interest. A chief's status depended on his statues: any chief who failed to cut trees to transport and erect statues would have found himself out of a job.

The Easter Islanders' isolation probably also explains why their collapse — more, perhaps, than the collapse of any other preindustrial society — haunts readers and visitors today. The parallels between Easter Island and the modern world are chillingly obvious. Thanks to globalization, international trade, jet planes, and the Internet, all countries on Earth today share resources and affect each other, just as did Easter's eleven clans. Polynesian Easter Island was as isolated in the Pacific Ocean as the Earth is today in space. When the Easter Islanders got into difficulties, there was nowhere to which they could flee or to which they could turn for help; nor shall we modern Earthlings have recourse elsewhere if our troubles increase. Those are the reasons why people see the collapse of Easter Island society as a metaphor, a worst-case scenario, for what may lie ahead of us in our own future.

Of the two new accounts of Easter Island's message that Flenley and Bahn and Van Tilburg have now given us, which would I recommend to readers? Both books are so interesting but so dissimilar that those of us attracted to history, exploration, and exotic societies will enjoy reading both. Those interested in none of those things but looking for a florid Gothic novel can read Van Tilburg's book and try to forget that it happens to be a true story.

JENNY EVERETT

My Little Brother on Drugs

FROM *Popular Science*

I SWIPE AN alcohol-soaked gauze pad over my younger brother's left thigh, an inch below the hem of his SpongeBob boxers. As I screw the needle into the injection pen, Alex feeds me instructions. It's my first time, but already it's his thirty-seventh.

"Here are the rules: insert the needle quickly and gently, but only when I say so," he says, taking the pen to pantomime the motion. He removes the first of two protective caps and turns a knob on the pen — one, two, three, four, five clicks — and watches intensely as his dose is released into the barrel.

"Make sure the skin dimples. That means the needle is all the way in," he continues. "Press the button until it clicks, then hold it there for five seconds. Keep the skin dimpled, otherwise all the medicine won't go in me. When you take out the needle, do it straight up and fast. And, Jenny, please don't hit a vein. That *huwts* me." Suddenly, dropping his *r*, Alex sounds much more like his nine-year-old self.

I pinch a clump of skin between my thumb and index finger and wait. "OK," he whispers. But I can't do it. *"OK,"* he repeats. I pierce the fatty tissue and wince — and take it as a compliment that he doesn't. "Keep dimpling!" he yells.

Here's the thing: my brother isn't sick. He's short. Shorter than every boy and girl in Ms. Lemcke's fourth-grade class, shorter than 97 percent of boys his age. What I've just shot into his 3-feet-11¾-inch, fifty-pound body is Humatrope, a lab-brewed human growth hormone (hGH) nearly identical to the hGH secreted by the pituitary gland, the critical metabolic hormone that regulates not only height, as its name suggests, but also cardiac function, fat metabolism, and muscle growth.

Alex's quest for "enheightenment," as I've come to call it, began last summer just as the Food and Drug Administration expanded its approved uses of Humatrope, Eli Lilly and Company's recombinant hGH, to include children of idiopathic short stature (ISS) — kids who are extremely short for reasons that are not entirely understood. Kids who, like Alex, are teased or ignored by classmates who may trump their height by a foot — but whose "condition" may be caused by nothing more than genetics. This groundbreaking and controversial FDA ruling made Humatrope available to 400,000 American children expected to grow no taller than 5 feet 3 in the case of boys and 4 feet 11 in the case of girls, putting them in the bottom 1.2 percent. For Alex, the nightly hGH shots will probably continue for six to eight years — all to make this otherwise healthy boy grow taller.

Human growth is an invisible but intense process, an intricate and little understood web of genes, hormones, and other variables. Genetics aside, growth hormone may be the single biggest player. Between ten and thirty times a day, your hypothalamus sends a growth hormone–releasing hormone to the garbanzo-bean-size pituitary gland at the base of your brain. Each time the pituitary gland receives a signal, it spits out a small amount of growth hormone. Although scientists think a small percentage of hGH travels to your bones, a majority of the hormone latches onto binding proteins, which carry it to receptors in your liver cells. This triggers the secretion of insulin-like growth factor-1 (IGF-1), a protein that promotes bone growth in children and teenagers until their growth plates, areas at the ends of the bones, fuse, at around age seventeen for boys or fifteen for girls. After that, growth hormone continues to regulate the metabolic system, burning fat and building muscle, but we produce exponentially less hGH each decade after puberty. Thus the teenager who can routinely "supersize it" without consequence ages into the thirty-year-old whose beer and burgers go straight to his gut.

In 1971, Berkeley chemist Choh Hao Li synthesized the growth hormone molecule, an enormous biotech breakthrough, and in 1985 synthetic growth hormone was approved by the FDA to treat growth hormone deficiency. Prior to the drug's development, medicinal growth hormone was scarce. What did exist had to be extracted from the pituitary glands of human corpses — most of the time legally, but occasionally by pathologists being paid by suppliers to remove the hormone without permission from the deceased's family. As a result of the shortage, hGH treatment was conservative, reserved for kids who made very little if any growth hormone themselves. Between 1963 and 1985, 7,700 patients in the United States took the hormone. Ultimately, 26 of these patients died of Creutzfeldt-Jakob disease (CJD); the fatal brain disease that is similar to mad cow is thought to have contaminated a batch of the pituitary hormone doled out in the sixties and seventies. The FDA banned growth hormone when the first two cases of CJD were reported in early 1985 — just in time for the agency's approval, later that year, of the synthetic version.

Suddenly, the growth hormone supply was unlimited, safer, and less expensive, opening the door for looser diagnoses, along with higher and more frequent doses. By 1988 what had once been a niche drug prescribed to treat dwarfism was shattering all market expectations, chalking up more than a half billion dollars in annual sales.

At the time, I was eleven years old and a ripe candidate for growth hormone supplementation. I had been born in the fifth percentile for height, but at age five my "growth velocity" started slipping, and by eleven, I was hovering below the first percentile. Blood tests revealed that I was producing only "borderline acceptable" levels of growth hormone. "Growth hormone injections may be an option," Yale pediatric endocrinologist William Tamborlane told my parents. The prospect terrified me. "I don't care if I'm a midget! It's what's inside that counts!" I protested. "I'm not having a shot every day."

Further testing showed that my thyroid gland was malfunctioning, a definable and common condition called hypothyroidism. Tamborlane prescribed a pill (which I will take every day for the rest of my life), and my growth velocity picked up immediately. Still, I was told that I would never surmount the magical 5-foot threshold.

That's because most short children can place primary blame for their stature on the genetic lottery. "Want to be taller, Jenny?" I remember Tamborlane asking me as his eyes shifted between my slumping growth chart, my X-rays, and my mom and dad. I nodded eagerly. "Well, you should've picked different parents." He chuckled, while I considered the merits of a parentectomy. Today I'm twenty-six years old and my height has topped off at . . . 5 feet 1, thank you very much.

Fifteen years after I flirted with growth hormone treatment, biotech's baby has exploded into a $1.5 billion industry that has reached more than 200,000 children and sent many more to the doctor wondering if hGH is for them — including my brother Alex. When my mom called to tell me Alex was growth-hormone-deficient and would soon begin injections, I was skeptical. The prospect of having a shot every day had sent fear through my own little body, and now, as an overprotective big sister, I didn't want my brother's carefree childhood to be interrupted by such stress — and such a serious, understudied medical treatment — unless it was entirely necessary. Is being short so horrible that it should be medicalized and treated as an illness?

My thoughts were interrupted by my mom's voice, bringing me back to reality. "Nurses are coming over next week to show us how to administer the injection," she said. The decision had been made.

As the no. 6 train squeals under Manhattan's Upper East Side, Alex's Nikes sway 3 inches above the floor. My brother is a happy kid whose sprite profile doesn't resemble that of the typical round-faced growth-hormone-deficient cherub. He's silent now, but only because his mouth is full of jellybeans. The train reaches our stop, and Alex places his sticky little hand in mine.

"Excuse me," he says politely as we jostle our way off.

A surprised middle-aged mom looks up from her book. "My, you are much more polite than *my* kindergartener," she says.

It's been about two weeks since Alex began his nightly injections. Before that, he might have flinched at this well-intentioned underestimate of his age. But today he squares himself a bit and responds with a certain pride: "I'm nine years old," he says, "and I'm on Humatrope!"

It strikes me, not for the first time, just how important the drug has become to Alex. He yearns to be taller. As the youngest of six, he knows how to get noticed — our family joke is that he swallowed an amplifier — and what he lacks in stature he was dealt in personality. But at school, where tall kids hold the social scepter, his big personality is overlooked.

"Everyone says that it's what's inside that counts, and that makes me feel good," Alex says, "but if I was the tallest instead of the shortest, everything would just be better. People would sit with me at lunch, I'd have more friends, and people in my class wouldn't make fun of me and call me Little Everett. And I'd be a better soccer goalie. I think 6 feet would be good."

Studies show that half of short kids report being teased and three-quarters say they're treated as younger than their age — but keep in mind that this is an awkward age to begin with. Starting at about age ten, grade-wide height discrepancies tend to widen dramatically, which just adds to the broader stew of middle-school insecurities that most of us shudder to recall. This is the age at which families often become conscious of and concerned about growth issues — because discrepancies are suddenly so visually apparent.

When I was in fifth grade, the boys would chase me down the hall on their knees playing "catch the midget," and one of the tougher girls would taunt me: "You're so small I need a microscope to see you." But before feeling sorry for me, consider my classmate Kelly (her name has been changed), who was 5 feet 10, a full 2 feet taller than I was — a virtual fifth-grade giant. (Given social norms, it's appropriate to compare the experience of being a short boy to that of being a tall girl.) Kelly was tortured so badly throughout adolescence that at eighteen she became bulimic. When her weight dropped to 115 pounds, she was hospitalized. "I thought it would make me smaller and more attractive," she recalled in a recent phone conversation. "It almost killed me."

Being a tall girl is so psychologically traumatic, in fact, that in the 1950s, doctors began giving tall girls estrogen as a growth suppressant. In high doses, the hormone stimulates cartilage maturation without causing an increase in height, which means the girls stop growing earlier. In Kelly's case, treatment was discussed, but doctors were confident, based on her bone development, that she wouldn't grow much taller than 6 feet. Good call: today Kelly is only an inch taller than she was in fifth grade. And although no formal long-term studies have been done, tall girls treated with estro-

gen have reported increased incidences of miscarriage, endome-
triosis, infertility, and ovarian cysts. Yet a survey taken last year
reported that one-third of pediatric endocrinologists have offered
the treatment at least once in the past five years.

Learning about the side effects of estrogen therapy only increased my apprehension about treating Alex with hGH. Armed
with a small forest of research, I drove up to Yale–New Haven
Children's Hospital to unload my worries on Dr. Myron Genel,
Alex's pediatric endocrinologist.

With his white hair and white lab coat, Genel looks like the proto-
typical old-school doctor, the kind you imagine making house calls
with his little black medical kit. I take a sip of my coffee. "This won't
stunt my growth, will it?" I joke. "Too late to worry about that," he
replies.

I reach into my bag and pull out binders of research: diagrams
of the growth process, historical time lines, and a folder labeled
"risks," which is packed with studies. "How concerned should my
family be about the risks?" I ask.

I wait for reassurance — or, in any case, for a defense of Alex's
treatment. But Genel surprises me.

"We honestly don't know the long-term side effects, and I think
that's a reason for real concern," he says. "We're using a hormone
that promotes growth, and there are things whose growth we don't
want to promote. IGF-1, for example, has been shown to play a role
in the development of malignancies in tissue culture."

This I knew. Multiple studies of human serum specimens have
shown that elevated levels of IGF-1 identify people at higher risk
for developing breast, prostate, and colon cancer, and tumor specimens, most studies show, have more IGF-1 receptors than normal
adjacent cells. Although it's not yet known whether an abundance
of IGF-1 actually causes malignancies or is merely associated with
some other risk factor, it is a reason for concern because growth
hormone is what stimulates the liver and tissues to produce IGF-1.

But Genel explains that he'll test Alex's IGF levels every three
months to make sure they're in what's considered the normal
range. "In theory, if we keep his growth factors at an acceptable
level, he's not at risk," he says.

I scan my list of questions, typed up in order of progressing intensity, and zoom to the bottom. "Given the risks, what makes Alex a good candidate?"

Again his response surprises me — this time because it challenges my assumption that Alex is in fact a good candidate for treatment. "We can define those youngsters who make virtually no growth hormone, because they have a very typical presentation," he says. "And we can generally sort out those youngsters who make an ample amount of growth hormone. We have a very difficult time, however, defining youngsters like your brother, who make some growth hormone, but who possibly don't make enough."

He explains that although Alex's levels of IGF-1 are low and his height has gradually declined to the first percentile, he does produce *some* growth hormone.

"Your brother is a murky case, and there are enough questions about the safety and efficacy of this drug that I cannot say one way or another whether he should definitely receive treatment," he continues. "Frankly, I felt we could wait — but not very much longer — and gather information. It was a decision that your family made and, I suspect, Alex made."

Here is a manifestation of just how complicated and unpredictable the growth process is: it's impossible to measure hGH levels using a simple blood test. Because the hormone oscillates in the blood, constantly peaking and sinking, a hundred samples can yield a hundred different answers. So doctors must rely on growth hormone stimulation tests, where the patient is injected with an artificial agent that stimulates the pituitary gland to produce growth hormone. After the agent is injected, nurses sample the patient's blood every thirty minutes for two hours, hoping to catch the pituitary operating at full strength.

"These are artificial tests," Genel says. "None of them tell us anything about what a youngster does under normal circumstances. It only tells us that if you give them an artificial stimulus, the pituitary gland will release hormone."

Genel shows me my brother's test results: Alex produced readings ranging from 0.11 to 9.9 ng/ml. Though most doctors look for a top level of at least 10 as an indicator of healthy hormone production, Yale has a lower benchmark of 7 or greater. So by Yale's standards, Alex passed. My parents, I realize, have no idea that Alex may not be growth-hormone-deficient by some careful definition;

they believe he is clinically deficient. Essentially, my little brother is an experiment.

Over dinner that night, my mother and father recall the day Alex's test results arrived in the mail. Startled by the low numbers, they assumed he was producing far too little growth hormone. They had no idea growth hormone production is so difficult to measure. "I'm a parent, not a scientist," my mom says. "I shouldn't have to know that." It's safe to take him off the drug, I tell them: "It's not too late to change your mind."

But despite the day's revelations, they decide to move forward with treatment. Alex's confidence has already soared since starting Humatrope; they don't have the heart to disappoint him. Besides, my parents say, they're concerned Alex will have fewer professional opportunities, and they worry he won't find a woman to be with, if his height remains in the basement.

Such anxieties, of course, are hardly unique to my parents — and, in fact, a quick glance at the research would appear to back them up. A recent University of Florida study, for example, found that each extra inch of height amounted to $789 more a year in pay. So someone who is 6 feet would be expected, on average, to earn $5,523 more annually than someone who is 5 feet 5. Another study indicates that just 3 percent of Fortune 500 CEOs are shorter than 5 feet 7, and more than half are taller than 6 feet, though only 20 percent of the population is.

But it's not as simple an equation as these numbers make it seem, says Dr. David Sandberg, an associate professor of psychiatry and pediatrics at the University of Buffalo. Sandberg's studies have found that although short kids are teased and treated as younger than their age, there is no evidence that making them 2½ inches taller will make any difference in their quality of life. "Our lives are so much more complicated than one single factor," Sandberg says.

In clinical trials by Humatrope manufacturer Eli Lilly, children taking the drug grew on average 1 to 1½ inches more than the placebo group; 62 percent of the kids tested grew more than 2 inches over their predicted adult height, and 31 percent gained more than 4 inches. This would land Alex, whose predicted height without growth hormone is about 5 feet 6, somewhere between 5 feet 7 and 5 feet 10.

Dr. Harvey Guyda, chair of the department of pediatrics at McGill University in Canada, questions the studies, especially given what he describes as a high dropout rate. In the Eli Lilly studies, he points out, only 28 percent of the placebo and 42 percent of the growth hormone–therapy subjects completed the study; it seems reasonable to assume, he says, that the subjects who endured the study were the ones who demonstrated the most extreme growth.

"The mantra is that healthy, short kids are handicapped, abnormal, have all sorts of problems, and we have to do something," says Guyda, who testified against the FDA's approval of Humatrope for treating kids with ISS. "But there is no data to prove that these kids are any different from normal-stature children, and there is absolutely zero data that says when you give growth hormone to a kid who's said to have this psychosocial problem because he's short there's any benefit. Prove to me that a few extra inches is worth the cost of daily injections."

Financially alone, that cost, according to Guyda, amounts to $10,000 per centimeter for a growth hormone–deficient kid, and somewhere between $22,000 and $43,000 per centimeter for kids with idiopathic short stature. For now, my parents' insurance company, Anthem Blue Cross and Blue Shield, has agreed to cover Alex's treatment, but not every short child has insurance or a pediatric endocrinologist to recommend treatment. The not-surprising result is that any advantage hGH does confer will likely go to already advantaged patients: rich, white American males. (For every two girls who receive treatment, five boys do; this is at least partly explained by the fact that boys suffer more discrimination in relation to their short stature than girls do.)

A week before Alex's three-month checkup, I return to Yale, this time to meet with Tamborlane, my own pediatric endocrinologist, whom I haven't seen since my last appointment eight years ago. I'm especially interested in his opinion about Alex because Tamborlane, now the chief of pediatric endocrinology at Yale, voted for the approval of Humatrope to treat idiopathic short stature. We meet in the cafeteria, and although I'm not looking for a free checkup, he palpates my throat right there. "Thyroid feels healthy," he reports.

Tamborlane voted for the approval, he tells me, because the drug was already approved to treat some groups of children who

produce plenty of growth hormone — children with Turner syndrome, a genetic abnormality; children born small for gestational
age; and children with chronic renal insufficiency, a kidney disease.
In each of these cases, hGH isn't treating the disease, it's treating
the resulting undesirable physical characteristic — short stature.
In kids with ISS, though, it's even more ambiguous because the disease, if there is one, is unknown.

I present Alex's case to Tamborlane, explaining my family's mo-
tivations and the uncertainty surrounding the diagnosis. I tell him
that although I want Alex to have every advantage and the best
possible quality of life, I'm concerned about the drug's unclear
benefits and the potential long-term risks for kids who are short for
reasons that aren't fully understood.

"If Alex were your son," I ask, "would you put him on hGH?"

Tamborlane leans back and pauses to consider the question.

"Given the uncertainties, probably not. I was a short, geeky kid at
a football prep school, and I survived — maybe even gained something from it," he says. (He's now 5 feet 9.) "All signs say Alex would
probably grow to a very livable height without the growth hormone."

In early November, we bring Alex to Yale for his first checkup. By
now the injection has become routine — slotted into a nine P.M.
Nickelodeon commercial break — and my family is already notic-
ing changes in Alex's body. His muscle tone is visibly improved,
and his pants are suddenly too short. In the waiting room, I jot
down a few things to mention to the doctor: Alex's appetite is un-
characteristically voracious; his growth pains are intensifying; and
his hair is dry and brittle.

The room used for measuring height and weight is wallpapered
with drawings by patients with a variety of metabolic disorders. One
depicts a small, sad-looking stick figure labeled "before" and a
taller, much happier stick figure labeled "after." Alex stands to the
left of this, his back to the wall, grinning in anticipation of his "after." The nurse scribbles on his chart and ushers us into the examining room. Alex is anxious because while we swear he's grown at
least 3 inches, we're not positive because we haven't measured him
yet — psychologists recommend we don't measure Alex at home.

Finally Genel announces, "124.8 cm. That's just about 4 feet 1,
about an inch and a half in three months."

"Wow," says my mom, extrapolating 6 inches of growth a year.

"That's *all?*" says Alex. "All those shots for one measly inch?!"

Genel warns them not to take the results too literally. "These are *positive* results," he says to Alex. Then, turning to my mom: "But it's too early to attribute this response to the treatment."

Indeed, there's no telling what the results of Alex's treatment will be; even at $20,000 a year there are no guarantees. Some kids end up in the sixtieth percentile, while others never crawl above the fifth. I'm still left wondering why we're rolling the dice on a healthy kid, particularly when the benefits of a few extra inches are unproven and the risks are unknown. At the same time, I'm rooting for results — and we leave the doctor's office cautiously optimistic that the treatment is having an impact. We're further reassured three months later, in early February, when Alex's next official measurement comes in: He's 4 feet 2½ inches, a full 2½ inches taller than he was six months earlier when he began treatment.

TIMOTHY FERRIS

Stumbling into Space

FROM *The New York Review of Books*

1

George W. Bush's January 14 speech at NASA headquarters, in which he set the manned space program on a new trajectory, was an oddly dissociated event. NASA administrator Sean O'Keefe stood alone at stage left with his arms hanging limply at his sides and his fingers curled, looking like an eagle that has just eaten a gratifyingly plump mouse but is having trouble digesting it. The president, adopting his customary tank-window squint, briefly praised shuttle astronauts for conducting "important research" and helping to build the International Space Station — and then enthused about the "stunning images" from NASA space telescopes and the investigations being conducted by its probes of Mars, Jupiter, and Saturn. The odd thing was that aside from Bush's tip of the hat to the shuttle and the station — whose death warrants he was signing — all the triumphs he cited were the work of *un*manned robotic spacecraft.

Which pretty much reflects the problem. NASA is two agencies — three, if you count aerospace — in one. Its unmanned programs are flying high: robotic probes have sampled the sands of Mars, mapped every planet in the solar system this side of Pluto, inspected comets and asteroids, photographed infant galaxies near the edge of the observable universe, and made incalculable contributions to terrestrial communications, agriculture, geology, and weather forecasting, all at a fraction of the cost of sending astronauts up there. Meanwhile, the manned program is stuck in low

Earth orbit. As Bush noted, "In the past thirty years, no human being has set foot on another world, or ventured farther upward into
space than 386 miles — roughly the distance from Washington,
D.C., to Boston, Massachusetts." It is as if sixteenth-century Spain,
three decades after Columbus, lacked a single ship capable of venturing out of sight of land.

Can the Bush plan get manned space flight going again — and
should it?

Bush was amply justified in deciding to retire the shuttle, which
despite all its merits — it is, after all, the world's only winged, reusable spaceship — never had a raison d'être and had become both
an emblem and a cause of NASA's woes. Dreamed up as a kind of
hangover cure in the days following the *Apollo* lunar missions, when
the NASA budget was shrinking from over 4 percent of the federal
budget to its current level of under 1 percent, the shuttle was sold
to Congress as a cost-effective way of putting humans and satellites
in orbit. Taken in by NASA hype, President Nixon assured the nation that "a space vehicle that can shuttle repeatedly from Earth to
orbit and back . . . will revolutionize transportation into near space,
by routinizing it," and President Reagan declared, following three
test flights of the first space shuttle, *Columbia,* that shuttles were
now "ready to provide economical and routine access to space."

This was sheer fantasy, as NASA was in a position to know and
ought to have admitted. Economical? The shuttle substantially
raised the costs of putting vehicles into orbit, rather than reducing
them: at well over $300 million per launch, it costs $10,000 per
pound for the shuttle to deliver a payload into orbit — two to
five times the going commercial rate. Routine? Combine two of
the space agency's own predictions — that the shuttle would fly almost weekly, which never happened, and that it would realize a 98
percent safety record, which turned out to be about right — and
you're looking at a shuttle crash every year. To have funded that
sort of prospective carnage indicates that neither NASA nor the
Nixon White House believed the forecasts on which they were bas-
ing their decisions.

The process of purveying such claims sent NASA down the slippery slope of believing its own press releases, a degeneration re-
marked upon by both the Rogers Commission, which investigated

the explosion of the shuttle *Challenger* 73 seconds after launch on January 28, 1986, and the *Columbia* Accident Investigation Board, convened after that shuttle disintegrated while reentering the atmosphere on February 1, 2003. The *Columbia* board, chaired by Harold Gehman, Jr. — a sixty-year-old retired four-star admiral given to the unrelenting pursuit of hard facts — has produced a clearer and more coherent report than did the Rogers Commission, but it makes an even more depressing read, since so little at NASA seems to have changed during the seventeen years separating the two accidents. It concludes that the *Columbia* crash was "not a random event, but rather a product of . . . a series of political compromises that produced unreasonable expectations — even myths — about its performance." NASA bought into these myths, at one point boasting that the shuttle was "the most reliable, flexible, and cost-effective launch system in the world" when in fact, according to the report, it was and still is "a developmental vehicle that operates not in routine flight but in the realm of dangerous exploration."

The proximate causes of *Columbia*'s demise were not particularly difficult to discern, although the Gehman board took pains to verify them and to analyze competing hypotheses. In essence, two things happened. First, 82 seconds after launch, a 2.7-pound chunk of insulating foam broke loose from the shuttle's fuel tank and knocked a hole in the leading edge of its left wing. Second, although NASA soon became aware of the incident, it took inadequate steps to determine the extent of the damage and did nothing at all to protect the astronauts against what proved to be its lethal effects.

At launch, the space shuttle system consists of a reusable spaceship, or "orbiter," about the size of a DC-9, attached to the side of a gigantic fuel tank, 154 feet long and 28 feet in diameter, containing the liquid hydrogen/oxygen fuel that powers the orbiter's engines. (Additional power is provided by a pair of solid rocket boosters, the malfunction of one of which blew up the *Challenger*.) Because hydrogen and oxygen have to be kept extremely cold in order to remain liquid rather than turn into vapor, the tank is covered with a layer of insulating foam. Keeping the foam glued firmly to the freezing, sweating tank had long been a headache for NASA engineers, and on many occasions pieces of it had come loose, hit the orbiter, and done damage. The worst such debris strike occurred during a launch in December 1988, when the shuttle *Atlantis* sustained a flabbergasting 707 foam hits. Mission commander R. L. "Hoot" Gibson inspected the damage with a video camera attached to the shuttle's robotic arm. (*Columbia* had no such arm.) "It looked like we had been shotgun-blasted," he recalled. "I looked at those pictures and said, 'We are going to die' to myself." Ground control pronounced the damage acceptable and *Atlantis* landed without incident. It turned out to have lost a heat-protection tile, but fortunately an aluminum plate happened to lie beneath the tile. Otherwise, in Gibson's estimation, his mission might not have made it home.

The very fact that so many shuttles survived foam strikes evidently lulled NASA into underestimating the danger. Although photos showing the debris strike during *Columbia*'s final launch raised concerns on the ground just hours into the mission, the Gehman board found that NASA "declined to have the crew inspect the orbiter for damage, declined to request on-orbit imaging, and ultimately discounted the possibility of a burn-through" — that is, that there might be a hole in the wing's thermal protection layer through which hot plasma, generated by friction when the shuttle reentered Earth's atmosphere, could invade the wing and blowtorch it from within. Unaware of any problem, the seven astronauts left orbit on schedule, seventeen days into the mission, cheerfully shooting a last few videos showing the red glow of plasma dancing outside the flight deck windows as they began their descent into the upper atmosphere. Soon thereafter the shuttle came apart, etching epitaphic skywriting across dawn skies from California to Louisiana. A forensic analysis conducted by a Crew Survivability Working Group convened at the Gehman board's request concluded that the module containing the crew remained intact for approximately 24 seconds after the orbiter broke up, during which time it fell from an altitude of approximately 140,000 feet to a little over 100,000 feet before disintegrating.

NASA's failure to order imaging of the shuttle to look for signs of wing damage while it was in orbit is as inexplicable as it is anguishing. The air force has a number of telescopic cameras on Earth that are capable of taking high-resolution pictures of an orbiting shuttle. The cameras' precise capacities are classified, but a conservative estimate, based on air force data, is that they make visible an object the size of a golf ball at the shuttle's distance. The Gehman board estimated that the hole in *Columbia*'s wing measured 100 square inches — about the size of a bucket of golf balls. Several shuttle ground team members suggested commissioning images by such cameras. One warned in an e-mail that he and his team would "always have big uncertainties" about the damage "until we get definitive, better, clearer photos of the wing and body underside," adding, "Can we petition (beg) for outside . . . assistance?"

The answer was no. Incredibly, it seems that nobody on hand appreciated what the cameras could do — although that much was common knowledge among amateur astronomers and space buffs — and efforts by engineers to find out by contacting the Defense Department directly were scotched by their superiors. On Flight Day 6, one of the NASA officials in charge of the mission, Linda Ham, told the mission management team, which she chaired, that on-orbit imaging was not being pursued because even if it revealed damage, "there is not much we can do about it." This was not the case; as we shall see, plenty could have been done to attempt a rescue of *Columbia*'s crew. But the fact that mission managers "displayed no interest in understanding [the] problem and its implications," as the Gehman board put it, seems to have resulted from NASA's prideful reluctance to ask for aid from other agencies, combined with its old habit of downplaying hazards it had survived in the past.

As the reentering shuttle passed over Kirtland Air Force Base in Albuquerque on the morning of February 1, scientists there took a few images of it for their own amusement, using an ancient Macintosh computer and a digital camera coupled to a 3.5-inch Questar — a portable telescope so small that it can be transported as carry-on luggage on a commercial flight. The equipment didn't work very well and clouds got in the way, but the scientists did manage to take one overexposed photograph when the shuttle was near the horizon. Even this crude image suggested that something was going terribly wrong, but by then it was of course too late. Meanwhile the advanced Starfire adaptive-optics telescopes at Kirtland stood idle, NASA officials having neglected to employ them. A week after the crash, Space Shuttle Program Manager Ron Dittemore displayed the Kirtland scientists' fuzzy photo to reporters at a press conference, offering it as an example of how ground-based imaging wouldn't have shown anything useful. "If your eye is sharp, maybe you can draw a conclusion," he said. "I don't think it's very revealing." Watching him on live TV, I was startled to realize that although Dittemore had managed eleven shuttle missions and was immersed in the dreadful business of trying to determine what had doomed the last one, he *still* didn't comprehend the contribution that ground-based imaging could have made to *Columbia*'s safety.

What might NASA have done to rescue the astronauts, had it recognized the peril they faced? *Columbia* could not have retreated to the International Space Station — it lacked sufficient fuel to move to the station's radically different orbit — but an analysis conducted at the Gehman board's request came up with a list of other options. Here is how things might have been:

1. Responding to launch-film images of the foam impact, NASA promptly obtains high-resolution ground-camera imagery of the orbiting shuttle. If the images show damage or are inconclusive, two *Columbia* crew members put on space suits and go out of the orbiter to get a firsthand look at the wing, taking pictures of it for analysis on the ground. (The strike occurred at a part of the wing too close to the shuttle fuselage for the astronauts on board to see it without first leaving the craft.) Having confirmed that there is a hole on the leading edge of the left wing, the crew adopts a minimum-exertion schedule, canceling the performance of nonessential tasks and sleeping as long as possible to prolong the time they can remain in orbit before their oxygen and CO_2-scrubbing supplies run out. These steps can extend the mission to Flight Day 30. It is now Flight Day 5.

2. The shuttle *Atlantis* is put on a rush flight-preparation schedule to launch a rescue mission with a skeleton crew consisting of a commander, pilot, and two astronauts trained in spacewalking. (Twenty-three such astronauts were available at the Kennedy Space Center.) They are advised of the situation, apprised of the danger that a second foam-related launch accident could doom their mission as well, and invited to volunteer. (How many would have volunteered? My guess is that all twenty-three would have volun-

teered.) NASA estimates that *Atlantis* could have launched at least five days before *Columbia*'s crew ran out of air.

3. In case *Atlantis* cannot reach them in time, members of *Columbia*'s crew again go out into space, cramming the hole in the wing with tools and pieces of metal and filling it with water, which freezes into ice in the cold vacuum of space. This step, in addition to jettisoning extra weight and adopting a sashay-style reentry flight path to minimize the heating of the left wing, might hold the shuttle together through thermal reentry. Then, at an altitude of around 30,000 feet, the crew could activate their emergency escape system and bail out — since the wing and left landing gear might not hold together on landing — leaving the empty shuttle to crash in a "disposal area" in the Pacific south of Fiji.

4. *Atlantis* launches safely, however, and docks within thirty feet of *Columbia*. Escorted by the two spacewalking astronauts, the *Columbia* crew crosses over to *Atlantis,* which returns safely to Earth with all eleven men and women aboard — a mission well within its capacities. Once they are back on Earth, *Columbia,* which cannot be landed by remote control, is deorbited and ditched. Seven lives have been saved, at an acceptable level of risk to four others.

NASA now says that when shuttle missions resume next year, a second shuttle will always be kept at the ready in case a rescue is required. That may save lives someday, but it does little to reduce the sorrow and shame one feels when considering that NASA could have pulled off the greatest rescue in the history of space flight rather than presiding over one of its worst disasters. The officials directly at fault deserve to be held responsible, and have been, but as the Gehman board notes, the crash resulted from "persistent, systemic flaws" in NASA management, and problems of this magnitude cannot be solved just by changing the nameplates on office doors. "Both accidents were 'failures of foresight' in which history played a prominent role," the board concluded. By "history" they meant the dangerous habit of becoming complacent about a persistent hazard. NASA managers knew they had problems with flying foam prior to the *Columbia* crash — just as, seventeen years earlier, they knew they had problems with the O-rings that held parts of the shuttle together, before an O-ring failure blew up the *Challenger.*

The board also recommended transformations in NASA's "culture," but conceded that "the changes we recommend will be difficult to accomplish — and will be internally resisted." In other words, don't hold your breath. If manned space flight is going to get substantially safer, the best hope is not to create a sociological revolution at NASA but to "replace the shuttle as soon as possible" with a new and safer vehicle. Bush didn't have much choice about axing the shuttle.

2

But replace it with what, and for what purpose? Bush's NASA speech set three goals: (1) get the space shuttle flying again and use it to finish construction of the International Space Station, nominally within the next six years; (2) replace the shuttle with a new manned spacecraft, called the Crew Exploration Vehicle, within ten years; the CEV will most likely be an *Apollo*-style capsule rather than a winged space plane; (3) use the new craft to commence "extended human missions to the moon as early as 2015, with the goal of living and working there for increasingly extended periods."

The first goal was uncontroversial for those who accept Bush's premises. From his point of view, the only real alternative was to halt manned space flight altogether before reconstituting it a decade hence for the moon missions, which might well cost more than keeping it going in the interim. But the first goal is also unpromising. The shuttle is already on its way out and the space station been pointless from the start. Notwithstanding a lot of talk about making perfectly spherical ball bearings in zero gravity aboard the station or "learning to live and work in space" by being there, the station has never done much except give the shuttle somewhere to go. (Bush pointedly avoided repeating the old hyperbole and instead simply said, bleakly enough, that station astronauts are to study "the long-term effects of space travel on human biology" — something the Russians already did, years ago.) So why pour further tens of billions of dollars into the station? For Bush and many others, national pride is in question, plus the fact that much of the money has already been committed. "We will finish what we have started, we will meet our obligations to our fifteen international partners on this project," Bush said. Translation: "We've bought too much dog food to shoot the dog now."

*

Promptly after Bush spoke, NASA canceled all future shuttle missions to the Hubble Space Telescope. Without hands-on servicing to replace its gyroscopes and update other equipment, the space telescope will become unusable by around 2007 and fall out of the sky by 2014. The reason cited by NASA was safety — if shuttle astronauts get in trouble while at Hubble, they cannot go to the International Space Station to await rescue. But flying dozens of missions to the space station is more dangerous than flying just one mission to Hubble, and the presence of a second rescue shuttle at Cape Canaveral further reduces the risk. Some of the shuttle astronauts' finest hours have been spent repairing and refurbishing Hubble, which ranks among the most productive and popular scientific instruments ever constructed.

Just as I was writing this paragraph, an e-mail came in announcing a new Hubble finding that dark energy, the mysterious antigravity force currently accelerating the cosmic expansion rate, probably isn't strong enough to tear the universe apart within the next 30 billion years. You don't have to be a cosmologist to consider that discerning the fate of the universe is more important than measuring bone marrow changes in space-station astronauts. We can hope that NASA can find the will and a way to save Hubble before it's too late. The National Academy of Sciences is now reviewing the situation, partly in response to a letter from Admiral Gehman arguing that "only a deep and rich study of the entire gain/risk equation can answer the question of whether an extension of the life of the wonderful Hubble telescope is worth the risks involved."

Bush's second goal, developing a safer spacecraft to replace the shuttle, makes sense if the new craft has a meaningful mission. Unfortunately its first assignment, as Bush put it, will be "ferrying astronauts and scientists to the Space Station after the shuttle is retired" — a prospect every bit as dismal as it sounds. We're talking tens of billions of dollars to staff the world's most expensive medical experiment.

Things get slightly brighter, however, with Bush's third proposal: to reopen the moon to human exploration. Retrograde as it may seem — haven't we already been to the moon? — this actually may not be such a bad idea.

If we take a long view, the ultimate goal of manned space exploration is to establish permanent homes for humanity elsewhere in the solar system. Centuries from now we might expect to find exploratory and scientific outposts scattered all the way from here to Saturn, with substantial numbers of people living on Mars and perhaps even Venus — assuming that both planets can be transformed ("terraformed") to endow them with oceans, breathable atmospheres, and abundant indigenous life.

There are good reasons to want to do this. In addition to providing *homo sapiens* with an insurance policy — a pan-planetary human species could survive terrestrial disasters, such as global warming or an asteroid impact, that could otherwise doom us — it would open up vast frontiers for exploration, habitation, and exploitation. Some of the more intriguing strategies for establishing a foothold on Mars call for sending explorers who, from the start, go there to stay: the growing Mars colony always keeps enough spacecraft on hand to serve as lifeboats if all or some of the settlers have to bail out, but the idea is to manufacture rocket fuel, grow crops, put down roots, and make a go of it. The moon is an excellent place to develop the technologies and skills required for such an effort. It's a harsh, airless world, to be sure — tougher than Mars in many respects — but as big as Africa and a lot cheaper to get to than Mars is.

Maintaining a permanent lunar base means, however, that somebody has to keep footing the bills. Unless taxpayers are ready to bear the entire burden, ways must be found to make money on the moon. Safe storage of electronic data is one near-future possibility: the moon will remain untouched by virtually any conceivable terrestrial catastrophe, and streams of electronic data going directly from rooftop antennas to a backup facility on the moon are almost impossible to intercept. Lunar/solar power is another prospect. Worldwide demand for electricity is expected to increase five to ten times by the middle of this century. Unless a clean new power source such as nuclear fusion is soon available, that means choosing between growing pollution, politically unpalatable nuclear fission reactors, or the unlikely spectacle of vast arrays of windmills and solar panels blanketing entire counties. The moon has plenty of empty land where solar panels could be deployed. Preliminary studies indicate that power collected by lunar arrays of such panels could safely be sent to terrestrial collecting stations as low-energy microwave beams. Once nuclear fusion becomes possible, moreover, helium-3, the isotope burned by the cleanest, safest kind of prospective fusion reactor, could be to the moon what gold was to California and oil to Pennsylvania and Texas: helium-3, rare on Earth, is abundant on the moon.

Tourism is another potential source of revenue, especially if lunar soil proves to contain water ice, which could be melted for drinking water and broken down into hydrogen and oxygen to make rocket fuel to ferry tourists from a low orbit around Earth up to the moon. (If the moon turns out to be bone-dry, nuclear rockets might serve this purpose, but bringing water up would cancel out much of the savings.) Meanwhile the high cost of getting from the Earth to low orbit could be avoided by building a space elevator — essentially a cable attached to the ground at the bottom and to a geosynchronous satellite at the top. The satellite would orbit at the same rate the Earth rotates (as the communications satellites that feed rooftop TV dishes do) so the cable would stay put. Electrical elevators would shuttle up and down the cable, carrying people and goods into orbit quickly and inexpensively.

Aside from the cable itself, which would have to be made of a material stronger and lighter than any currently available (carbon nanotubes are promising candidates, as are noncarbon tubes based on tungsten and molybdenum), startup costs are the main obstacle to building a space elevator. You need a big, heavy satellite with a mighty power-generating station. Nobody today could afford to haul such a thing up from Earth, but if the heavy material were brought down from the moon, the project could become profitable. And once a single space elevator was installed, the cost of going anywhere else in the solar system, whether to mine asteroids for precious minerals or to farm Mars, would be drastically reduced.

In this sense the moon really could be the gateway to the great beyond. But making money, whether by mining lunar fuel or enticing tourists to zero-gravity space hotels, is the kind of activity best pursued by the private sector, not the federal government — and as Greg Klerkx notes in *Lost in Space: The Fall of NASA and the Dream of a New Space Age*, NASA has long turned a cold shoulder to private-sector space enthusiasts with profit-making ideas. "The idea of space as an individual frontier [is] anathema to NASA's way of doing business," he writes. "When it comes to actually engaging in commercial enterprises, NASA has failed resoundingly." He quotes Alan Ladwig, a perspicacious NASA veteran, as charging that the space agency views entrepreneurs with deep suspicion. "There were plenty of things you could do from a commercial point of view, but they might end up taking a NASA person's job," Ladwig told Klerkx. "That was the perspective." So long as a lunar base remains exclusively a NASA operation, it is likely to be about as innovative and interesting — and as profitable — as NASA food.

If you take human space exploration seriously, the stage would seem to be set for genuine collaboration between government and private enterprise, in something like the way that the Northwest Ordinance of 1787, the Homestead Act of 1862, and concessions to the railroad and telegraph industries opened up the American West. Raw meat for a Republican administration, one would think, yet Bush has done nothing to welcome entrepreneurs with a bent for working in space — the sort of people who in prior generations took wagon trains west, wildcatted for oil, or designed and flew early aircraft, and who now are looking upward to the final frontier. Instead his plan, worked up by NASA administrator O'Keefe and the National Security Council with little or no input from the private sector, amounts to more of the same old model, in which space is explored exclusively by federal employees who wear stars and stripes on their sleeves, spending billions and earning next to nothing.

As it has in many other policy matters, White House silence on this score leaves the public guessing whether the administration has no new ideas or prefers not to talk about them. My guess is the former. Bush seems to care little about space exploration generally or his own plan in particular. He didn't even mention it in his State of the Union address, and in any case he knows that it will take years to get going, by which time even a two-term Bush would be out of office.

If indeed we're in for more of the same, the future of manned space flight looks dim. Without the involvement of private visionaries and entrepreneurs, shuttle missions to nowhere will drag on for years, to be followed by an expensive lunar base that threatens to drain resources from NASA's useful (and blameless) unmanned projects. We could of course abandon manned space exploration altogether, but the cure might be worse than the disease. The NASA budget would most likely shrink to half its current size — anyone who thinks Congress would shift all that money to space science is dreaming — with the surplus going not to relieve hunger or illiteracy but to pay a few weeks' interest on the national debt.

The question is one of acting sensibly in the short term while keeping our eyes on the potentialities of the future. Trying to plan today how to settle on Mars makes about as much sense as asking Columbus to come up with a way to bring water to Los Angeles. Yet if we cease exploring until we know all the answers, that day will never come. The alternative to blundering ahead with manned space flight is that either some other nation or group of nations will eventually colonize the moon and Mars, or nobody will. Either way, as Walt Kelly's Porkypine mused in another context, it's a mighty sobering thought.

MALCOLM GLADWELL

Getting Over It

FROM *The New Yorker*

WHEN TOM RATH, the hero of Sloan Wilson's 1955 novel, *The Man in the Gray Flannel Suit,* comes home to Connecticut each day from his job in Manhattan, his wife mixes him a martini. If he misses the train, he'll duck into the bar at Grand Central Terminal and have a highball, or perhaps a Scotch. On Sunday mornings, Rath and his wife lie around drinking martinis. Once, Rath takes a tumbler of martinis to bed, and after finishing it drifts off to sleep. Then his wife wakes him up in the middle of the night, wanting to talk. "I will if you get me a drink," he says. She comes back with a glass half full of ice and gin. "On Greentree Avenue cocktail parties started at seven-thirty, when the men came home from New York, and they usually continued without any dinner until three or four o'clock in the morning," Wilson writes of the tidy neighborhood in Westport where Rath and countless other young, middle-class families live. "Somewhere around nine-thirty in the evening, martinis and Man-hattans would give way to highballs, but the formality of eating any-thing but hors d'oeuvres in-between had been entirely omitted."

The Man in the Gray Flannel Suit is about a public relations special-ist who lives in the suburbs, works for a media company in mid-town, and worries about money, job security, and educating his children. It was an enormous bestseller. Gregory Peck played Tom Rath in the Hollywood version, and today, on the eve of the fiftieth anniversary of the book's publication, many of the themes the novel addresses seem strikingly contemporary. But in other ways *The Man in the Gray Flannel Suit* is utterly dated. The details are all wrong. Tom Rath, despite an introspective streak, is supposed to be

a figure of middle-class normalcy. But by our standards he and almost everyone else in the novel look like alcoholics. The book is
supposed to be an argument for the importance of family over career. But Rath's three children — the objects of his sacrifice — are
so absent from the narrative and from Rath's consciousness that
these days he'd be called an absentee father.

The most discordant note, though, is struck by the account of
Rath's experience in World War II. He had, it becomes clear, a terrible war. As a paratrooper in Europe, he and his close friend Hank
Mahoney find themselves trapped — starving and freezing — behind enemy lines, and end up killing two German sentries in order
to take their sheepskin coats. But Rath doesn't quite kill one of
them, and Mahoney urges him to finish the job:

> Tom had knelt beside the sentry. He had not thought it would be dif-
> ficult, but the tendons of the boy's neck had proved tough, and sud-
> denly the sentry had started to sit up. In a rage Tom had plunged the
> knife repeatedly into his throat, ramming it home with all his strength
> until he had almost severed the head from the body.

At the end of the war, Rath and Mahoney are transferred to
the Pacific theater for the invasion of the island of Karkow. There
Rath throws a hand grenade and inadvertently kills his friend. He
crawls over to Hank's body, calling out his name. "Tom had put his
hand under Mahoney's arm and turned him over," Wilson writes.
"Mahoney's entire chest had been torn away, leaving the naked
lungs and splintered ribs exposed."

Rath picks up the body and runs back toward his own men,
dodging enemy fire. Coming upon a group of Japanese firing from
a cave, he props the body up, crawls within fifteen feet of the machine gun, tosses in two grenades, and then finishes off the lone
survivor with a knife. He takes Hank's body into a bombed-out pill-
box and tries to resuscitate his friend's corpse. The medics tell him
that Hank has been dead for hours. He won't listen. In a daze, he
runs with the body toward the sea.

Wilson's description of Mahoney's death is as brutal and moving
a description of the madness of combat as can be found in postwar
fiction. But what happens to Rath as a result of that day in Karkow?
Not much. It does not destroy him or leave him permanently
traumatized. The part of Rath's war experience that leaves him

truly guilt-ridden is the adulterous affair that he has with a woman named Maria while waiting for redeployment orders in Rome. In the elevator of his midtown office, he runs into a friend who knew Maria and learns that he fathered a son. He obsessively goes over and over the affair in his mind, trying to square his feeling toward Maria with his love for his wife, and his marriage is fully restored only when he confesses to the existence of his Italian child. Killing his best friend, by contrast, is something that comes up and then gets tucked away. As Rath sat on the beach, and Mahoney's body was finally taken away, Wilson writes:

> A major, coming to squat beside him, said, "Some of these goddamn sailors got heads. They went ashore and got Jap heads, and they tried to boil them in the galley to get the skulls for souvenirs."
>
> Tom had shrugged and said nothing. The fact that he had been too quick to throw a hand grenade and had killed Mahoney, the fact that some young sailors had wanted skulls for souvenirs, and the fact that a few hundred men had lost their lives to take the island of Karkow — all these facts were simply incomprehensible and had to be forgotten. That, he had decided, was the final truth of the war, and he had greeted it with relief, greeted it eagerly, the simple fact that it was incomprehensible and had to be forgotten. Things just happen, he had decided; they happen and they happen again, and anybody who tries to make sense out of it goes out of his mind.

You couldn't write that scene today, at least not without irony. No soldier, according to our contemporary understanding, could ever shrug off an experience like that. Today, it is Rath's affair with Maria that would be rationalized and explained away. He was a soldier, after all, in the midst of war. Who knew if he would ever see his wife again? Tim O'Brien's best-selling 1994 novel *In the Lake of the Woods* has a narrative structure almost identical to that of *The Man in the Gray Flannel Suit*. O'Brien's hero, John Wade, is present at a massacre of civilians in the Vietnamese village of Thuan Yen. He kills a fellow soldier — a man he loved like a brother. And, just like Rath, Wade sits down at the end of the long afternoon of the worst day of his war and tries to wish the memory away:

> And then later still, snagged in the sunlight, he gave himself over to forgetfulness. "Go away," he murmured. He waited a moment, then said it again, firmly, much louder, and the little village began to vanish inside

its own rosy glow. Here, he reasoned, was the most majestic trick of all. In the months and years ahead, John Wade would remember Thuan Yen the way chemical nightmares are remembered, impossible combinations, impossible events, and over time the impossibility itself would become the richest and deepest and most profound memory.

This could not have happened. Therefore it did not.

Already he felt better.

But John Wade cannot forget. That's the point of O'Brien's book. *The Man in the Gray Flannel Suit* ends with Tom Rath stronger and his marriage renewed. Wade falls apart, and when he returns home to the woman he left behind he wakes up screaming in his sleep. By the end of the novel, the past has come back and destroyed Wade, and one reason for the book's power is the inevitability of that disaster. This is the difference between a novel written in the middle of the last century and a novel written at the end of the century. Somehow in the intervening decades our understanding of what it means to experience a traumatic event has changed. We believe in John Wade now, not Tom Rath, and half a century after the publication of *The Man in the Gray Flannel Suit* it's worth wondering whether we've got it right.

Several years ago, three psychologists — Bruce Rind, Robert Bauserman, and Philip Tromovitch — published an article on childhood sexual abuse in *Psychological Bulletin,* one of academic psychology's most prestigious journals. It was what psychologists call a meta-analysis. The three researchers collected fifty-nine studies that had been conducted over the years on the long-term psychological effects of childhood sexual abuse (CSA) and combined the data in order to get the most definitive and statistically powerful result possible.

What most studies of sexual abuse show is that if you gauge the psychological health of young adults — typically college students — using various measures of mental health (alcohol problems, depression, anxiety, eating disorders, obsessive-compulsive symptoms, social adjustment, sleeping problems, suicidal thoughts and behavior, and so on), those with a history of childhood sexual abuse will have more problems across the board than those who weren't abused. That makes intuitive sense. But Rind and his colleagues wanted to answer that question more specifically: how *much* worse off were the sexually abused? The fifty-nine studies were run through a series of sophisticated statistical tests. Studies from different times and places were put on the same scale. The results were surprising. The difference between the psychological health of those who had been abused and those who hadn't, they found, was marginal. It was two-tenths of a standard deviation. "That's like the difference between someone with an IQ of 100 and someone with an IQ of 97," Rind says. "Ninety-seven is statistically different from 100. But it's a trivial difference."

Then Rind and his colleagues went one step further. A significant percentage of people who were sexually abused as children grew up in families with a host of other problems, like violence, neglect, and verbal abuse. So, to the extent that the sexually abused were damaged, what caused the damage — the sexual abuse, or the violence and neglect that so often accompanied the abuse? The data suggested that it was the latter, and, if you account for such factors, that two-tenths of a standard deviation shrinks even more. "The real gap is probably smaller than 100 and 97," Rind says. "It might be 98, or maybe it's 99." The studies analyzed by Rind and his colleagues show that some victims of sexual abuse don't even regard themselves, in retrospect, as victims. Among the male college students surveyed, for instance, Rind and his colleagues found that "37 percent viewed their CSA experiences as positive at the time they occurred," while 42 percent viewed them as positive when reflecting back on them.

The Rind article was published in the summer of 1998, and almost immediately it was denounced by conservative groups and lambasted in the media. Laura Schlessinger — a popular radio talk-show host known as Dr. Laura — called it "junk science." In Washington, Representative Matt Salmon called it "the Emancipation Proclamation for pedophiles," while Representative Tom DeLay accused it of "normalizing pedophilia." They held a press conference at which they demanded that the American Psychological Association censure the paper. In July of 1999, a year after its publication, both the House and the Senate overwhelmingly passed resolutions condemning the analysis. Few articles in the history of academic psychology have created such a stir.

But why? It's not as if the authors said that CSA was a good thing. They just suggested that it didn't cause as many problems as we'd thought — and the question of whether CSA is morally wrong doesn't hinge on its long-term consequences. Nor did the study say that sexual abuse was harmless. On *average*, the researchers concluded, the long-term damage is small. But that average is made up of cases where the damage is hard to find (like CSA involving adolescent boys) and cases where the damage is quite significant (like father-daughter incest). Rind was trying to help psychologists focus on what was truly harmful. And, when it came to the effects of things like physical abuse and neglect, he and his colleagues sounded the alarm. "What happens in physical abuse is that it doesn't happen once," Rind says. "It happens time and time again. And, when it comes to neglect, the research shows that is the most noxious factor of all — worse than physical abuse. Why? Because it's not practiced for one week. It's a persistent thing. It's a permanent feature of the parent-child relationship. These are the kinds of things that cause problems in adulthood."

All Rind and his colleagues were saying is that sexual abuse is often something that people eventually can get over, and one of the reasons that the Rind study was so unacceptable is that we no longer think that traumatic experiences are things we can get over. We believe that the child who is molested by an uncle or a priest, on two or three furtive occasions, has to be permanently scarred by the experience — just as the soldier who accidentally kills his best friend must do more than sit down on the beach and decide that sometimes things just "happen."

In a recent history of the Rind controversy, the psychologist Scott Lilienfeld pointed out that when we find out that something we thought was very dangerous actually isn't that dangerous after all we usually regard what we've learned as good news. To him, the controversy was a paradox, and he is quite right. This attachment we have to John Wade over Tom Rath is not merely a preference for one kind of war narrative over another. It is a shift in perception so profound that the United States Congress could be presented with evidence of the unexpected strength and resilience of the human spirit and reject it without a single dissenting vote.

In *The Man in the Gray Flannel Suit*, Tom Rath works for Ralph Hopkins, who is the president of the United Broadcasting Company. Hopkins has decided that he wants to play a civic role in the issue of mental health, and Rath's job is to write his speeches and handle public relations connected to the project. "It all started when a group of doctors called on me a few months ago," Hopkins tells Rath, when he hires him for the job. "They apparently felt that there is too little public understanding of the whole question of mental illness, and that a campaign like the fight against cancer or polio is needed." Again and again in the novel, the topic of mental health surfaces. Rath's father, we learn, suffered a nervous breakdown after serving in the trenches of World War I and died in what may well have been a suicide. His grandmother, whose death sets the book's plot in motion, wanders in and out of lucidity at the end of her life. Hopkins, in a hilarious scene, recalls his unsatisfactory experience with a psychiatrist. To Wilson's readers, this preoccupation would not have seemed out of place. In 1955 the population of New York State's twenty-seven psychiatric hospitals was nearly 94,000. (Today, largely because of antipsychotic drugs, it is less than 6,000.) It was impossible to drive any distance from Manhattan and not be confronted with ominous, hulking reminders of psychiatric distress: the enormous complex across the Triborough Bridge, on Wards Island; Sagamore and Pilgrim Hospitals, on Long Island; Creedmoor, in Queens. Mental health mattered to the reader of the 1950s in a way that, say, AIDS mattered in the novels of the late 1980s.

But Wilson draws a very clear line between the struggles of the Raths and the plight of those suffering from actual mental illness. At one point, for example, Rath's wife, Betsy, wonders why nothing is fun anymore:

> It probably would take a psychiatrist to answer that. Maybe Tom and I both ought to visit one, she thought. What's the matter? the psychiatrist would say, and I would reply, I don't know — nothing seems to be much fun any more. All of a sudden the music stopped, and it didn't start again. Is that strange, or does it happen to everyone about the time when youth starts to go?
>
> The psychiatrist would have an explanation, Betsy thought, but I don't want to hear it. People rely too much on explanations these days, and not enough on courage and action . . . Tom has a good job, and he'll get his enthusiasm back, be a success at it. Everything's going to be fine. It does no good to wallow in night thoughts. In God we trust, and that's that.

This is not denial, much as it may sound like it. Betsy Rath is not saying that her husband doesn't have problems. She's just saying that, in all likelihood, Tom will get over his problems. This is precisely the idea that lies at the heart of the Rind meta-analysis. Once you've separated out the small number of seriously damaged people — the victims of father-daughter incest or of prolonged neglect and physical abuse — the balance of CSA survivors are pretty much going to be fine. The same is true, it turns out, of other kinds of trauma. The Columbia University psychologist George Bonanno, for instance, followed a large number of men and women who had recently lost a spouse. "In the bereavement area, the assumption has been that when people lose a loved one there is a kind of unitary process that everybody must go through," Bonanno says. "That process has been called grief work. The grief must be processed. It must be examined. It must be fully understood, then finished. It was the same kind of assumption that dominated the trauma world. The idea was that everybody exposed to these kinds of events will have to go through the same kind of process if they are to recover. And if you don't do this, if you have somehow inhibited or buried the experience, the assumption was that you would pay in the long run."

Instead, Bonanno found a wide range of responses. Some people went through a long and painful grieving process; others a period of debilitating depression. But by far the most common response was resilience: the majority of those who had just suffered from one of the most painful experiences of their lives never lapsed into serious depression, experienced a relatively brief period of grief symptoms, and soon returned to normal functioning. These people were not necessarily the hardiest or the healthiest. They just managed, by one means or another, to muddle through.

"Most people just plain cope well," Bonanno says. "The vast majority of people get over traumatic events, and get over them remarkably well. Only a small subset — 5 to 15 percent — struggle in a way that says they need help."

What these patterns of resilience suggest is that human beings are naturally endowed with a kind of psychological immune system, which keeps us in balance and overcomes wild swings to either end of the emotional spectrum. Most of us aren't resilient just in the wake of bad experiences, after all. We're also resilient in the wake of wonderful experiences; the joy of a really good meal, or winning a tennis match, or getting praised by a boss doesn't last that long, either. "One function of emotions is to signal to people quickly which things in their environments are dangerous and should be avoided and which are positive and should be approached," Timothy Wilson, a psychologist at the University of Virginia, has said. "People have very fast emotional reactions to events that serve as signals, informing them what to do. A problem with prolonged emotional reactions to past events is that it might be more difficult for these signals to get through. If people are still in a state of bliss over yesterday's success, today's dangers and hazards might be more difficult to recognize." (Wilson, incidentally, is Sloan Wilson's nephew.)

Wilson and his longtime collaborator, Daniel T. Gilbert, argue that a distinctive feature of this resilience is that people don't realize that they possess it. People are bad at forecasting their emotions — at appreciating how well, under most circumstances, they will recover. Not long ago, for instance, Gilbert, Wilson, and two other researchers — Carey Morewedge and Jane Risen — asked passengers at a subway station in Cambridge, Massachusetts, how much regret they thought they would feel if they arrived on the platform just as a train was pulling away. Then they approached passengers who really had arrived just as their train was leaving, and asked them how they felt. They found that the predictions of how bad it would feel to have just barely missed a train were on average greater than reports of how it actually felt to watch the train pull away. We suffer from what Wilson and Gilbert call an impact bias: we always assume that our emotional states will last much longer than they do. We forget that other experiences will compete for our attention and emotions. We forget that our psychological immune system will kick in and take away the sting of adversity. "When I talk about our research, I say to people, 'I'm not telling you that bad things don't hurt,'" Gilbert says. "Of course they do. It would be perverse to say that having a child or a spouse die is not a big deal. All I'm saying is that the reality doesn't meet the expectation."

This is the difference between our own era and the one of half a century ago — between *The Man in the Gray Flannel Suit* and *In the Lake of the Woods*. Sloan Wilson's book came from a time and a culture that had the confidence and wisdom to understand this truth.

"I love you more than I can tell," Rath says to his wife at the end of the novel. It's an ending that no one would write today, but only because we have become blind to the fact that the past — in all but the worst of cases — sooner or later fades away. Betsy turns back to her husband:

> "I want you to be able to talk to me about the war. It might help us to understand each other. Did you really kill seventeen men?"
>
> "Yes."
>
> "Do you want to talk about it now?"
>
> "No. It's not that I want to and can't — it's just that I'd rather think about the future. About getting a new car and driving up to Vermont with you tomorrow."
>
> "That will be fun. It's not an insane world. At least, our part of it doesn't have to be."

Personality Plus

FROM *The New Yorker*

WHEN Alexander (Sandy) Nininger was twenty-three and newly commissioned as a lieutenant in the United States Army, he was sent to the South Pacific to serve with the 57th Infantry of the Philippine Scouts. It was January 1942. The Japanese had just seized Philippine ports at Vigan, Legazpi, Lamon Bay, and Lingayen, and forced the American and Philippine forces to retreat into Bataan, a rugged peninsula on the South China Sea. There, besieged and outnumbered, the Americans set to work building a defensive line, digging foxholes and constructing dikes and clearing underbrush to provide unobstructed sight lines for rifles and machine guns. Nininger's men were on the line's right flank. They labored day and night. The heat and the mosquitoes were nearly unbearable.

Quiet by nature, Nininger was tall and slender, with wavy blond hair. As Franklin M. Reck recounts in *Beyond the Call of Duty*, Nininger had graduated near the top of his class at West Point, where he chaired the lecture and entertainment committee. He had spent many hours with a friend, discussing everything from history to the theory of relativity. He loved the theater. In the evenings, he could often be found sitting by the fireplace in the living room of his commanding officer, sipping tea and listening to Tchaikovsky. As a boy, he once saw his father kill a hawk and had been repulsed. When he went into active service, he wrote a friend to say that he had no feelings of hate and did not think he could ever kill anyone out of hatred. He had none of the swagger of the natural warrior. He worked hard and had a strong sense of duty.

In the second week of January, the Japanese attacked, slipping hundreds of snipers through the American lines, climbing into trees, turning the battlefield into what Reck calls a "gigantic possum hunt." On the morning of January 12, Nininger went to his commanding officer. He wanted, he said, to be assigned to another company, one that was in the thick of the action, so he could go hunting for Japanese snipers.

He took several grenades and ammunition belts, slung a Garand rifle over his shoulder, and grabbed a submachine gun. Starting at the point where the fighting was heaviest — near the position of the battalion's K Company — he crawled through the jungle and shot a Japanese soldier out of a tree. He shot and killed snipers. He threw grenades into enemy positions. He was wounded in the leg, but he kept going, clearing out Japanese positions for the other members of K Company, behind him. He soon ran out of grenades and switched to his rifle, and then, when he ran out of ammunition, used only his bayonet. He was wounded a second time, but when a medic crawled toward him to help bring him back behind the lines, Nininger waved him off. He saw a Japanese bunker up ahead. As he leaped out of a shell hole, he was spun around by a bullet to the shoulder, but he kept charging at the bunker, where a Japanese officer and two enlisted men were dug in. He dispatched one soldier with a double thrust of his bayonet, clubbed down the other, and bayoneted the officer. Then, with outstretched arms, he collapsed face down. For his heroism, Nininger was posthumously awarded the Medal of Honor, the first American soldier so decorated in World War II.

Suppose that you were a senior army officer in the early days of World War II and were trying to put together a crack team of fearless and ferocious fighters. Sandy Nininger, it now appears, had exactly the right kind of personality for that assignment, but is there any way you could have known this beforehand? It clearly wouldn't have helped to ask Nininger if he was fearless and ferocious, because he didn't know that he was fearless and ferocious. Nor would it have worked to talk to people who had spent time with him. His friend would have told you only that Nininger was quiet and thoughtful and loved the theater, and his commanding officer would have talked about the evenings of tea and Tchaikovsky. With the exception, perhaps, of the Scarlet Pimpernel, a love of music, theater, and long afternoons in front of a teapot is not a known predictor of great valor. What you need is some kind of sophisticated psychological instrument, capable of getting to the heart of his personality.

Over the course of the past century, psychology has been consumed with the search for this kind of magical instrument. Hermann Rorschach proposed that great meaning lay in the way that people described inkblots. The creators of the Minnesota Multiphasic Personality Inventory believed in the revelatory power of true-false items such as "I have never had any black, tarry-looking bowel movements" or "If the money were right, I would like to work for a circus or a carnival." Today, Annie Murphy Paul tells us in her fascinating new book, *Cult of Personality,* that there are 2,500 kinds of personality tests. Testing is a $400-million-a-year industry. A hefty percentage of American corporations use personality tests as part of the hiring and promotion process. The tests figure in custody battles and in sentencing and parole decisions. "Yet despite their prevalence — and the importance of the matters they are called upon to decide — personality tests have received surprisingly little scrutiny," Paul writes. We can call in the psychologists. We can give Sandy Nininger a battery of tests. But will any of it help?

One of the most popular personality tests in the world is the Myers-Briggs Type Indicator (MBTI), a psychological assessment system based on Carl Jung's notion that people make sense of the world through a series of psychological frames. Some people are extroverts, some are introverts. Some process information through logical thought. Some are directed by their feelings. Some make sense of the world through intuitive leaps. Others collect data through their senses. To these three categories — (I)ntroversion/(E)xtroversion, i(N)tuition/(S)ensing, (T)hinking/(F)eeling — the Myers-Briggs test adds a fourth: (J)udging/(P)erceiving. Judgers "like to live in a planned, orderly way, seeking to regulate and manage their lives," according to an MBTI guide, whereas Perceivers "like to live in a flexible, spontaneous way, seeking to experience and understand life, rather than control it." The MBTI asks the test taker to answer a series of "forced-choice" questions, where one choice identifies you as belonging to one of these paired traits. The basic test takes twenty minutes, and at the end you are presented with a precise, multidimensional summary of your personality —

your type might be INTJ or ESFP, or some other combination. Two and a half million Americans a year take the Myers-Briggs. Eighty-nine companies out of the Fortune 100 make use of it, for things like hiring or training sessions to help employees "understand" themselves or their colleagues. Annie Murphy Paul says that at the eminent consulting firm McKinsey, "'associates' often know their colleagues' four-letter MBTI types by heart," the way they might know their own weight or (this being McKinsey) their SAT scores.

It is tempting to think, then, that we could figure out the Myers-Briggs type that corresponds best to commando work, and then test to see whether Sandy Nininger fits the profile. Unfortunately, the notion of personality type is not nearly as straightforward as it appears. For example, the Myers-Briggs poses a series of items grouped around the issue of whether you — the test taker — are someone who likes to plan your day or evening beforehand or someone who prefers to be spontaneous. The idea is obviously to determine whether you belong to the Judger or Perceiver camp, but the basic question here is surprisingly hard to answer. I think I'm someone who likes to be spontaneous. On the other hand, I have embarked on too many spontaneous evenings that ended up with my friends and me standing on the sidewalk, looking at each other and wondering what to do next. So I guess I'm a spontaneous person who recognizes that life usually goes more smoothly if I plan first, or, rather, I'm a person who prefers to be spontaneous only if there's someone around me who isn't. Does that make me spontaneous or not? I'm not sure. I suppose it means that I'm somewhere in the middle.

This is the first problem with the Myers-Briggs. It assumes that we are either one thing or another — Intuitive or Sensing, Introverted or Extroverted. But personality doesn't fit into neat binary categories; we fall somewhere along a continuum.

Here's another question:

> Would you rather work under a boss (or a teacher) who is good-natured but often inconsistent, or sharp-tongued but always logical?

On the Myers-Briggs, this is one of a series of questions intended to establish whether you are a Thinker or a Feeler. But I'm not sure I know how to answer this one either. I once had a good-natured boss whose inconsistency bothered me, because he exerted a great deal of day-to-day control over my work. Then I had a boss who was quite consistent and very sharp-tongued — but at that point I was in a job where day-to-day dealings with my boss were minimal, so his sharp tongue didn't matter that much. So what do I want in a boss? As far as I can tell, the only plausible answer is: it depends. The Myers-Briggs assumes that who we are is consistent from one situation to another. But surely what we want in a boss, and how we behave toward our boss, is affected by what kind of job we have.

This is the gist of the now famous critique that the psychologist Walter Mischel has made of personality testing. One of Mischel's studies involved watching children interact with one another at a summer camp. Aggressiveness was among the traits that he was interested in, so he watched the children in five different situations: how they behaved when approached by a peer, when teased by a peer, when praised by an adult, when punished by an adult, and when warned by an adult. He found that how aggressively a child responded in one of those situations wasn't a good predictor of how that same child responded in another situation. Just because a boy was aggressive in the face of being teased by another boy didn't mean that he would be aggressive in the face of being warned by an adult. On the other hand, if a child responded aggressively to being teased by a peer one day, it was a pretty good indicator that he'd respond aggressively to being teased by a peer the next day. We have a personality in the sense that we have a consistent pattern of behavior. But that pattern is complex and that personality is contingent: it represents an interaction between our internal disposition and tendencies and the situations that we find ourselves in.

It's not surprising, then, that the Myers-Briggs has a large problem with consistency: according to some studies, more than half of those who take the test a second time end up with a score that is different from when they took it the first time. Since personality is continuous, not dichotomous, clearly some people who are borderline Introverts or Feelers one week slide over to Extroversion or Thinking the next week. And since personality is contingent, not stable, how we answer is affected by which circumstances are foremost in our minds when we take the test. If I happen to remember my first boss, then I come out as a Thinker. If my mind is on my second boss, I come out as a Feeler. When I took the Myers-Briggs, I scored as an INTJ. But, if odds are that I'm going to be something else if I take the test again, what good is it?

Once, for fun, a friend and I devised our own personality test.

Like the MBTI, it has four dimensions. The first is Canine/Feline. In romantic relationships, are you the pursuer, who runs happily to the door, tail wagging? Or are you the pursued? The second is More/Different. Is it your intellectual style to gather and master as much information as you can or to make imaginative use of a discrete amount of information? The third is Insider/Outsider. Do you get along with your parents or do you define yourself outside your relationship with your mother and father? And, finally, there is Nibbler/Gobbler. Do you work steadily, in small increments, or do everything at once, in a big gulp? I'm quite pleased with the personality inventory we devised. It directly touches on four aspects of life and temperament — romance, cognition, family, and work style — that are only hinted at by Myers-Briggs. And it can be completed in under a minute, nineteen minutes faster than Myers-Briggs, an advantage not to be dismissed in today's fast-paced business environment. Of course, the four traits it measures are utterly arbitrary, based on what my friend and I came up with over the course of a phone call. But then again surely *all* universal dichotomous typing systems are arbitrary.

Where did the Myers-Briggs come from, after all? As Paul tells us, it began with a housewife from Washington, D.C., named Katharine Briggs, at the turn of the last century. Briggs had a daughter, Isabel, an only child for whom (as one relative put it) she did "everything but breathe." When Isabel was still in her teens, Katharine wrote a book-length manuscript about her daughter's remarkable childhood, calling her a "genius" and "a little Shakespeare." When Isabel went off to Swarthmore College, in 1915, the two exchanged letters nearly every day. Then one day Isabel brought home her college boyfriend and announced that they were to be married. His name was Clarence (Chief) Myers. He was tall and handsome and studying to be a lawyer, and he could not have been more different from the Briggs women. Katharine and Isabel were bold and imaginative and intuitive. Myers was practical and logical and detail oriented. Katharine could not understand her future son-in-law. "When the blissful young couple returned to Swarthmore," Paul writes, "Katharine retreated to her study, intent on 'figuring out Chief.'"

She began to read widely in psychology and philosophy. Then in 1923, she came across the first English translation of Carl Jung's *Psychological Types*. "This is it!" Katharine told her daughter. Paul recounts, "In a dramatic display of conviction she burned all her own research and adopted Jung's book as her 'Bible,' as she gushed in a letter to the man himself. His system explained it all: Lyman [Katharine's husband], Katharine, Isabel, and Chief were introverts; the two men were thinkers, while the women were feelers; and of course the Briggses were intuitives, while Chief was a senser." Encouraged by her mother, Isabel — who was living in Swarthmore and writing mystery novels — devised a paper-and-pencil test to help people identify which of the Jungian categories they belonged to, and then spent the rest of her life tirelessly and brilliantly promoting her creation.

The problem, as Paul points out, is that Myers and her mother did not actually understand Jung at all. Jung didn't believe that types were easily identifiable, and he didn't believe that people could be permanently slotted into one category or another. "Every individual is an exception to the rule," he wrote; to "stick labels on people at first sight," in his view, was "nothing but a childish parlor game." Why is a parlor game based on my desire to entertain my friends any less valid than a parlor game based on Katharine Briggs's obsession with her son-in-law?

The problems with the Myers-Briggs suggest that we need a test that is responsive to the complexity and variability of the human personality. And that is why, not long ago, I found myself in the office of a psychologist from New Jersey named Lon Gieser. He is among the country's leading experts on what is called the Thematic Apperception Test (TAT), an assessment tool developed in the 1930s by Henry Murray, one of the most influential psychologists of the twentieth century.

I sat in a chair facing Gieser, as if I were his patient. He had in his hand two dozen or so pictures — mostly black-and-white drawings — on legal-sized cards, all of which had been chosen by Murray years before. "These pictures present a series of scenes," Gieser said to me. "What I want you to do with each scene is tell a story with a beginning, a middle, and an end." He handed me the first card. It was of a young boy looking at a violin. I had imagined, as Gieser was describing the test to me, that it would be hard to come up with stories to match the pictures. As I quickly discovered, though, the exercise was relatively effortless: the stories just tumbled out.

"This is a young boy," I began.

His parents want him to take up the violin, and they've been encouraging him. I think he is uncertain whether he wants to be a violin player, and maybe even resents the imposition of having to play this instrument, which doesn't seem to have any appeal for him. He's not excited or thrilled about this. He'd rather be somewhere else. He's just sitting there looking at it, and dreading having to fulfill this parental obligation.

I continued in that vein for a few more minutes. Gieser gave me another card, this one of a muscular man clinging to a rope and looking off into the distance. "He's climbing up, not climbing down," I said, and went on:

It's out in public. It's some kind of big square, in Europe, and there is some kind of spectacle going on. It's the seventeenth or eighteenth century. The king is coming by in a carriage, and this man is shimmying up, so he can see over everyone else and get a better view of the king. I don't get the sense that he's any kind of highborn person. I think he aspires to be more than he is. And he's kind of getting a glimpse of the king as a way of giving himself a sense of what he could be, or what his own future could be like.

We went on like this for the better part of an hour, as I responded to twelve cards — each of people in various kinds of ambiguous situations. One picture showed a woman slumped on the ground with some small object next to her; another showed an attractive couple in a kind of angry embrace, apparently having an argument. (I said that the fight they were having was staged, that each was simply playing a role.) As I talked, Gieser took notes. Later he called me and gave me his impressions. "What came out was the way you deal with emotion," he said. "Even when you recognized the emotion, you distanced yourself from it. The underlying motive is this desire to avoid conflict. The other thing is that when there are opportunities to go to someone else and work stuff out, your character is always going off alone. There is a real avoidance of emotion and dealing with other people, and everyone goes to their own corners and works things out on their own."

How could Gieser make such a confident reading of my personality after listening to me for such a short time? I was baffled by this at first, because I felt that I had told a series of random and idiosyncratic stories. When I listened to the tape I had made of the session, though, I saw what Gieser had picked up on: my stories were exceedingly repetitive in just the way that he had identified. The final card that Gieser gave me was blank, and he asked me to imagine my own picture and tell a story about it. For some reason, what came to mind was Andrew Wyeth's famous painting *Christina's World,* of a woman alone in a field, her hair being blown by the wind. She was from the city, I said, and had come home to see her family in the country: "I think she is taking a walk. She is pondering some piece of important news. She has gone off from the rest of the people to think about it." Only later did I realize that in the actual painting the woman is not strolling through the field. She is crawling, desperately, on her hands and knees. How obvious could my aversion to strong emotion be?

The TAT has a number of cards that are used to assess achievement — that is, how interested someone is in getting ahead and succeeding in life. One is the card of the man on the rope; another is the boy looking at his violin. Gieser, in listening to my stories, concluded that I was very low in achievement:

> Some people say this kid is dreaming about being a great violinist, and he's going to make it. With you, it wasn't what he wanted to do at all. His parents were making him do it. With the rope climbing, some people do this Tarzan thing. They climb the pole and get to the top and feel this great achievement. You have him going up the rope — and why is he feeling the pleasure? Because he's seeing the king. He's still a nobody in the public square, looking at the king.

Now this is a little strange. I consider myself quite ambitious. On a questionnaire, if you asked me to rank how important getting ahead and being successful was to me, I'd check the "very important" box. But Gieser is suggesting that the TAT allowed him to glimpse another dimension of my personality.

This idea — that our personality can hold contradictory elements — is at the heart of *Strangers to Ourselves,* by the social psychologist Timothy D. Wilson. He is one of the discipline's most prominent researchers, and his book is what popular psychology ought to be (and rarely is): thoughtful, beautifully written, and full of unexpected insights. Wilson's interest is in what he calls the "adaptive unconscious" (not to be confused with the Freudian unconscious). The adaptive unconscious, in Wilson's description, is a

big computer in our brain which sits below the surface and evalu-
ates, filters, and looks for patterns in the mountain of data that
come in through our senses. That system, Wilson argues, has a personality: it has a set of patterns and responses and tendencies that
are laid down by our genes and our early childhood experiences.
These patterns are stable and hard to change, and we are only
dimly aware of them. On top of that, in his schema we have another
personality: it's the conscious identity that we create for ourselves
with the choices we make, the stories we tell about ourselves, and
the formal reasons we come up with to explain our motives and
feelings. Yet this "constructed self" has no particular connection
with the personality of our adaptive unconscious. In fact, they
could easily be at odds. Wilson writes:

> The adaptive unconscious is more likely to influence people's uncon-
> trolled, implicit responses, whereas the constructed self is more likely to
> influence people's deliberative, explicit responses. For example, the
> quick, spontaneous decision of whether to argue with a co-worker is
> likely to be under the control of one's nonconscious needs for power
> and affiliation. A more thoughtful decision about whether to invite a co-
> worker over for dinner is more likely to be under the control of one's
> conscious, self-attributed motives.

When Gieser said that he thought I was low in achievement,
then, he presumably saw in my stories an unconscious ambivalence
toward success. The TAT, he believes, allowed him to go beyond
the way I viewed myself and arrive at a reading with greater depth
and nuance.

Even if he's right, though, does this help us pick commandos?
I'm not so sure. Clearly, underneath Sandy Nininger's peaceful
façade there was another Nininger capable of great bravery and ferocity, and a TAT of Nininger might have given us a glimpse of that
part of who he was. But let's not forget that he volunteered for the
front lines: he made a conscious decision to put himself in the
heat of the action. What we really need is an understanding of
how those two sides of his personality interact in critical situations.
When is Sandy Nininger's commitment to peacefulness more, or
less, important than some unconscious ferocity?

The other problem with the TAT, of course, is that it's a subjective instrument. You could say that my story about the man climb-

ing the rope is evidence that I'm low in achievement or you could say that it shows a strong desire for social mobility. The climber wants to look down — not up — at the king in order to get a sense "of what he could be." You could say that my interpretation that the couple's fighting was staged was evidence of my aversion to strong emotion. Or you could say that it was evidence of my delight in deception and role-playing. This isn't to question Gieser's skill or experience as a diagnostician. The TAT is supposed to do no more than identify themes and problem areas, and I'm sure Gieser would be happy to put me on the couch for a year to explore those themes and see which of his initial hypotheses had any validity. But the reason employers want a magical instrument for measuring personality is that they don't have a year to work through the ambiguities. They need an answer now.

A larger limitation of both Myers-Briggs and the TAT is that they are indirect. Tests of this kind require us first to identify a personality trait that corresponds to the behavior we're interested in, and then to figure out how to measure that trait — but by then we're two steps removed from what we're after. And each of those steps represents an opportunity for error and distortion. Shouldn't we try, instead, to test directly for the behavior we're interested in? This is the idea that lies behind what's known as the Assessment Center, and the leading practitioner of this approach is a company called Development Dimensions International, or DDI. Companies trying to evaluate job applicants send them to DDI's headquarters, outside Pittsburgh, where they spend the day role-playing as business executives. When I contacted DDI, I was told that I was going to be Terry Turner, the head of the robotics division of a company called Global Solutions.

I arrived early in the morning and was led to an office. On the desk was a computer, a phone, and a tape recorder. In the corner of the room was a video camera, and on my desk was an agenda for the day. I had a long telephone conversation with a business partner from France. There were labor difficulties at an overseas plant. A new product — a robot for the home — had run into a series of technical glitches. I answered e-mails. I prepared and recorded a talk for a product-launch meeting. I gave a live interview to a local television reporter. In the afternoon, I met with another senior Global Solutions manager and presented a strategic plan for the future of the robotics division. It was a long, demanding day at the office, and when I left, a team of DDI specialists combed through copies of my e-mails, the audiotapes of my phone calls and my speech, and the videotapes of my interviews, and analyzed me across four dimensions: interpersonal skills, leadership skills, business management skills, and personal attributes. A few weeks later, I was given my report. Some of it was positive: I was a quick learner. I had good ideas. I expressed myself well, and — I was relieved to hear — wrote clearly. But, as the assessment of my performance made plain, I was something less than top management material:

> Although you did a remarkable job addressing matters, you tended to handle issues from a fairly lofty perch, pitching good ideas somewhat unilaterally while lobbing supporting rationale down to the team below . . . Had you brought your team closer to decisions by vesting them with greater accountability, responsibility and decision-making authority, they would have undoubtedly felt more engaged, satisfied and valued . . .
>
> In a somewhat similar vein, but on a slightly more interpersonal level, while you seemed to recognize the value of collaboration and building positive working relationships with people, you tended to take a purely businesslike approach to forging partnerships. You spoke of win/win solutions from a business perspective and your rationale for partnering and collaboration seemed to be based solely on business logic. Additionally, at times you did not respond to some of the softer, subtler cues that spoke to people's real frustrations, more personal feelings, or true point of view.

Ouch! Of course, when the DDI analysts said that I did not respond to "some of the softer, subtler cues that spoke to people's real frustrations, more personal feelings, or true point of view," they didn't mean that I was an insensitive person. They meant that I was insensitive in the role of manager. The TAT and MBTI aimed to make global assessments of the different aspects of my personality. My day as Terry Turner was meant to find out only what I'm like when I'm the head of the robotics division of Global Solutions. That's an important difference. It respects the role of situation and contingency in personality. It sidesteps the difficulty of integrating my unconscious self with my constructed self by looking at the way that my various selves interact in the real world. Most important, it offers the hope that with experience and attention I can construct a more appropriate executive "self." The Assessment Center is probably the best method that employers have for evaluating personality.

But could an Assessment Center help us identify the Sandy Niningers of the world? The center makes a behavioral prediction, and, as solid and specific as that prediction is, people are least predictable at those critical moments when prediction would be most valuable. The answer to the question whether my Terry Turner would be a good executive is, once again: it depends. It depends on what kind of company Global Solutions is, and on what kind of respect my coworkers have for me, and on how quickly I manage to correct my shortcomings, and on all kinds of other things that cannot be anticipated. The quality of being a good manager is, in the end, as irreducible as the quality of being a good friend. We think that a friend has to be loyal and nice and interesting — and that's certainly a good start. But people whom *we* don't find loyal, nice, or interesting have friends, too, because loyalty, niceness, and interestingness are *emergent* traits. They arise out of the interaction of two people, and all we really mean when we say that someone is interesting or nice is that they are interesting or nice *to us*.

All these difficulties do not mean that we should give up on the task of trying to understand and categorize one another. We could certainly send Sandy Nininger to an Assessment Center and find out whether, in a make-believe battle, he plays the role of commando with verve and discipline. We could talk to his friends and discover his love of music and theater. We could find out how he responded to the picture of the man on a rope. We could sit him down and have him do the Myers-Briggs and dutifully note that he is an Introverted, Intuitive, Thinking Judger, and, for good measure, take an extra minute to run him through my own favorite personality inventory and type him as a Canine, Different, Insider Gobbler. We will know all kinds of things about him then. His personnel file will be as thick as a phone book, and we can consult our findings whenever we make decisions about his future. We just have to acknowledge that his file will tell us little about the thing we're most interested in. For that, we have to join him in the jungles of Bataan.

JEROME GROOPMAN

The Grief Industry

FROM *The New Yorker*

Soon after the collapse of the World Trade Center, experts predicted that one out of five New Yorkers — some one and a half million people — would be traumatized by the tragedy and require psychological care. Within weeks, several thousand grief and crisis counselors arrived in the city. Some were dispatched by charitable and religious organizations; many others worked for private companies that provide services to businesses following catastrophes.

In the United States, grief and crisis counselors generally use a method called critical incident stress debriefing, which was created in 1974 by Jeffrey T. Mitchell, a Maryland paramedic who was studying for a master's degree in psychology. Mitchell had seen a gruesome accident while on the job: a young bride, still in her wedding dress, had been impaled when the car that her drunk husband was driving rear-ended a pickup truck loaded with pipes. He was unable to shake the memory. Six months later, he confided his troubles to a friend — a firefighter who had witnessed similar horrors. The friend asked him to describe exactly what he had seen. Mitchell felt greatly relieved by this conversation, and became convinced that he had stumbled across an invaluable therapeutic approach. Indeed, he came to think that if a "debriefing" conversation was held soon after an upsetting event it could help prevent the onset of posttraumatic stress disorder.

In 1983, Mitchell received a Ph.D. in human development, and he began crafting a structured seven-step debriefing regimen that could be applied to groups of paramedics, firefighters, and other professionals who regularly witnessed traumatic events. Six years later, he started a nonprofit organization, the International Critical Incident Stress Foundation, to teach debriefing and related methods. The foundation has grown steadily, and more than thirty thousand counselors are trained by it each year.

In a typical debriefing session, crisis counselors introduce themselves and provide basic information about common stress reactions — sleeplessness, headache, irritability — as well as more debilitating symptoms, like flashbacks and delusions. Each participant is then asked to identify himself, pinpoint where he was during the tragic event (or "critical incident"), and describe what he witnessed. This is known as the "fact phase." The discussion next turns in a more emotional direction, as each participant is asked to divulge what he was thinking during the event. The purpose of sharing such memories is, in part, to draw out group members who "bottle up" their emotions. At the end of this process, the conversation enters the "feeling phase," focusing on each participant's current reaction to the catastrophe. (The counselors ask questions like "What was the worst part of the incident for you personally?") Finally, the counselors discuss strategies for coping with stress and suggest services that can provide additional help; by the end of the session, participants are considered ready for "reëntry" into the world. The group does not meet for a follow-up session.

I recently spoke with a man who worked at a travel agency on Liberty Street, across from where the Twin Towers once stood. He had been in the subway when the towers collapsed, but after considerable difficulty he made it home safely. "I was called by the company the next day and told to report to headquarters on Thursday," he told me. His parent corporation, which was situated in midtown and had numerous offices throughout the city, had hired an organization called National Employee Assistance Providers to give debriefing sessions. Many of its counselors used texts created by Mitchell's foundation during their training.

Most debriefings occur between twelve and seventy-two hours after a catastrophe, according to *Blindsided: A Manager's Guide to Catastrophic Incidents in the Workplace,* by Bruce T. Blythe, the CEO of Crisis Management International, a company that offers psychological services. Blythe writes, "Earlier than that, people are likely too numbed to put their personal reactions into words; after seventy-two hours, people typically begin to 'seal over' emotionally." This "sealing over" is seen as dangerously "laying the ground" for PTSD. In most circumstances, employees are required to attend a debriefing session. Blythe writes, "Experience has shown that if attendance is voluntary, those most in need of support will not come, out of fear or discomfort."

The travel agent sat in a conference room with coworkers from the Liberty Street branch who had witnessed the collapse of the World Trade Center and had been evacuated from the building. Also attending the session were employees from uptown offices who had not witnessed the collapse or been at risk. In all, there were between twenty and thirty participants at this debriefing session. "There were two counselors, a man and a woman, and they encouraged us to tell our stories and vent our feelings," the travel agent told me.

When it was the agent's turn, he revealed to the group that at the time of the attacks he had been sitting in a subway car, just short of the Fulton Street station. The train came to an abrupt halt, the air conditioning went off, and the conductor announced that the train's doors were stuck. Passengers managed to pry open the doors; as they stepped onto the platform, a tremendous blast of black smoke filled the air. It blew a woman walking in front of the agent off her feet. He ran away from the billowing smoke and soon found himself pressed up against a turnstile exit that wouldn't budge. The crowd pushed behind him, and he began to struggle for air. ("I said to myself, 'I'm not dying here,'" he told the group.) He broke free of the mob and found a stairwell; when he arrived at street level, the air was so dark with soot that he still felt as if he were trapped underground. He walked north and eventually got home.

"I told what happened to me, and people started crying," he recalled. A colleague said she had made her way to the pier where she usually catches a ferry to her home in New Jersey. "She told everyone how she came across a dazed coworker walking aimlessly in the darkness, and how they both saw people jumping into the water even though there was no boat there," he said. Another employee from the Liberty Street branch spoke vividly about watching bodies fall from the towers.

I asked the agent whether he had chosen to attend the debriefing. "Well, they felt everyone should participate," he said. When he was asked if it had been helpful, he shrugged and said that, like most of his Liberty Street colleagues, he was relatively numb during the debriefing. "Some people burst into tears," he said. "But the people who were really crying hadn't even been downtown."

At the end of the session, the two counselors gave telephone numbers to the workers and encouraged them to call if they felt distressed. The travel agent had nightmares for weeks after the debriefing and often felt as if he were choking. Images similar to the ones he had described during the session would flash through his mind. He didn't pursue further therapy, though. "I had to take care of my family; they rely on me," he explained. After several months, he said, the flashbacks and the sense of choking subsided. "You just block it out," he said. "You have to get on with life."

The director of human resources at the travel agent's company told me that she had arranged the debriefing session because "it made me feel that I was doing something for the employees." She went on, "I saw behavior that worried me, people very upset after the attacks. I didn't want the company to seem unfeeling." Another concern that leads companies to hire debriefing services is the fear of litigation. Employees who have experienced a traumatic incident on the job, and who have subsequently been sidelined by PTSD, have sued their companies. The Web site for National Employee Assistance Providers claims that its debriefing program ensures "that the productivity of the work unit is not impaired."

Hundreds of similar debriefing sessions took place in Manhattan in the days following the September 11 attacks. Did they help? One debriefing company told me that 99.7 percent of the participants found the sessions beneficial. But such evaluations are subjective, and hardly scientific. In fact, only in the past few years has debriefing undergone serious scrutiny. Brett Litz, a research psychologist at Boston Veterans Affairs Medical Center who specializes in posttraumatic stress disorder, recently completed a randomized clinical trial of group debriefing of soldiers who were stationed in Kosovo. (Peacekeeping forces there were exposed to sniper fire and mine explosions, and discovered mass graves.) He summarized the academic verdict on debriefing as follows: "The techniques practiced by most American grief counselors to prevent PTSD are inert."

Clinical trials of individual psychological debriefings versus no intervention after a major trauma, such as a fire or a motor vehicle accident, have had discouraging results. Some researchers have claimed that debriefing can actually impede recovery. One study of burn victims, for example, found that patients who received debriefing were much more likely to report PTSD symptoms than patients in a control group. It may be that debriefing, by encouraging patients to open their wounds at a vulnerable moment, augments distress rather than lessens it.

Mitchell, the movement's founder, told me that debriefing has been "distorted and misapplied" by some private companies, and noted that some negative findings stem from studies of these unorthodox variants. His technique, he added, is meant only for "homogeneous groups who have had the same exposure to the same traumatic event," and sometimes crisis counselors brought together people who had experienced unrelated traumas. With firefighters who had, say, all watched one of their colleagues die, Mitchell said that his method had a "proven" beneficial effect. He could cite no rigorous clinical trials, however, in support of this claim.

Scientific studies suggest that, after a catastrophic event, most people are resilient and will recover spontaneously over time. A small percentage of individuals do not rebound, however, and require extended psychological care. The single intervention of a debriefing session does nothing to alter this consistent dynamic.

Despite the influx of counselors into Manhattan, most New Yorkers received no therapy following the attacks. Furthermore, data from surveys taken after September 11 contradicted the early predictions that there would be widespread psychological damage. A telephone survey of 988 adults living below 110th Street, conducted in October and November of 2001, found that only 7.5 percent had been diagnosed as having PTSD. (According to the American Psychiatric Association, a patient is said to have PTSD if, for a month or more after a tragic event, he experiences several of the classic symptoms: flashbacks, intrusive thoughts, and nightmares; avoidance of activities and places that are reminiscent of the trauma; emotional numbness; chronic insomnia.) A follow-up of this survey, in March 2002, found that only 1.7 percent of New

Yorkers suffered from prolonged PTSD. This finding indicates that the debriefing industry is predicated on a false notion: that we are all at high risk for PTSD after exposure to a traumatic event.

In the wake of a catastrophe like September 11, Litz told me, victims should not be asked to disclose their personal feelings about the event. All that is needed is "psychological first aid": victims should be taken to a safe place, given food and water, and provided with information about the status of friends and family. None of this, he added, requires the presence of a trained psychologist.

In 1917, a traumatic event on a scale similar to that of the September 11 attacks took place in Halifax, Nova Scotia. Two ships collided near the dock, one of which was carrying explosives and benzene, a flammable liquid. The crew abandoned this ship, and it drifted to the dock, where it exploded and destroyed the entire north end of the city — an area encompassing two and a half square miles. More than two thousand inhabitants were killed, and nine thousand were injured — many of them blinded and dismembered. The night after the explosion, a blizzard descended on Halifax, hindering the relief effort, and many people whose homes had been destroyed froze to death.

April Naturale is a psychiatric social worker who heads Project Liberty, a government-sponsored program that was established to coordinate the therapeutic response to September 11. Not long ago, she went to Halifax to read archival materials on the 1917 accident. "Some of those who survived seemed psychotic, hallucinating for days," she told me. One woman continued to speak solicitously to someone named Alma — her dead child; other victims were in such a state of shock that doctors were able to perform surgery on them without using chloroform. But after a week or so these disturbing symptoms spontaneously subsided in the vast majority of cases. These accounts led Naturale to conclude that psychiatric intervention in the wake of such an event should be minimal; the mind should be given time to heal itself. In short, the "abnormal" behavior witnessed in the aftermath of the explosion was actually part of a healthy process of recovery.

Malachy Corrigan, the director of the Counseling Service Unit of the New York City Fire Department, was once a proponent of debriefing — but months before the September 11 attacks he decided that it was generally not a beneficial technique. "Sometimes when we put people in a group and debriefed them, we gave them memories that they didn't have," he told me. "We didn't push them to psychosis or anything, but, because these guys were so close and they were all at the fire, they eventually convinced themselves that they did see something or did smell something when in fact they didn't." For the workers in the pit at Ground Zero, Corrigan enlisted other firefighters to be "peer counselors" and to provide moral support and educational information about the possible mental health impact of sustained trauma.

"It was like one huge extended family," Corrigan recalled. "We gave them a lot of information about PTSD, as well as about the burden that they would be putting on their own families. We quite boldly spoke about alcohol and drugs. And we focused on the anger that comes with grief, because the members were more than happy to display those symptoms. You are speaking their language when you talk about alcohol and anger. The simpler you keep the mental health concepts, the easier it is to engage them."

Naturale sees the approach that Corrigan took, with peers providing basic comforts, as the paradigm for civilians as well as for rescue workers. "Non–mental health professionals do not pathologize," she said. "They don't know the terminology, they don't know how to diagnose. The most helpful approach is to employ a public health model, using people in the community who aren't diagnosing you."

Scientists are now trying to determine what causes some people to fall victim to PTSD after a traumatic event like the September 11 attacks. Rachel Yehuda, a neuroscientist at the Bronx Veterans Affairs Medical Center, has studied both combat veterans and Holocaust survivors and has found that people with PTSD have significantly lower baseline levels of cortisol, a hormone that is released in the body during moments of stress. Cortisol, Yehuda theorizes, acts as a counterbalance to adrenaline, which is thought to play a role in the "imprinting" of horrific and intrusive memories. She speculates that the lack of cortisol allows adrenaline to act unopposed, so to speak — and this contributes to the development of PTSD.

Vulnerability to PTSD, Yehuda added, also depends in part on the intensity and duration of the trauma. Someone who witnessed the fall of the towers from afar is not as likely to develop the disorder as someone who worked on the fiftieth floor of Tower One and only narrowly escaped. An injury can also help precipitate PTSD, and the disorder is more likely to affect a civilian bystander than someone who is trained to face dangerous situations, like a police officer. A study performed thirty-four months after the Oklahoma City bombing found that the rate of PTSD was 23 percent among male civilian victims and only 13 percent among firefighters.

Other studies have found that people who are at greatest risk for PTSD have a history of childhood abuse, family dysfunction, or a preexisting psychological disorder. In order to properly combat PTSD, Yehuda told me, we need to have a baseline mental health profile on everyone. "Why don't we have a doctor check our stress level?" she asked. "Just like doctors check our cholesterol."

A 1996 study of American pilots who were prisoners of war in North Vietnam underscores the importance of baseline mental health. Although the pilots endured years of torture and, in many cases, solitary confinement, they showed a very low incidence of PTSD — presumably because pilots are screened for psychological health and trained for high-stress combat.

Although there are no published studies on PTSD among rescue workers at Ground Zero, Corrigan, who has assessed many of these individuals, says it is relatively low. He estimates that, of about fifteen thousand firefighters and emergency personnel, fewer than a hundred have developed full-blown PTSD. "There were a lot of therapy experts here in New York who were quite happy to tell everyone that firefighters would have PTSD," he told me. "But these folks have tremendous resiliency. People say firefighters are crazy to put themselves at risk, but they are mentally very healthy. They can sustain enormous amounts of stress and continue to function."

Some of the most promising treatment interventions for people with PTSD have been developed by Edna Foa, a professor of psychology at the University of Pennsylvania. Twenty years ago, she began a research project involving rape victims in the Philadelphia area. "Most women recover," Foa told me. "Only about 15 percent will develop PTSD symptoms." For these women, Foa devised a technique to "restore resilience," based on cognitive behavioral therapy. The victim is slowly taught to restructure her reactions to her memories of the rape. First, a therapist sits with the woman and asks her to close her eyes and recount the event in detail. (Unlike group debriefing, this takes place months after the event and is performed one on one.) Then the woman is told to repeat the story. Subsequent therapy sessions span some thirty to forty-five minutes each and are taped so that the rape victim can listen to them at home. "The story changes as it is relived," Foa told me. "It becomes more organized, more flowing. A narrative emerges, with a beginning, a middle, and an end."

In contrast to classical psychotherapy, which attempts to link the patient's current feelings and behavior to previous events, Foa's treatment is focused primarily on relieving symptoms of distress. After each session, the patient is given homework assignments that are simple and direct. She is instructed to make a list of "avoidance behaviors," such as not getting into an elevator because it reminds her of the scene of her violation, and to record how anxious she feels when she listens to the tape or thinks about the rape. The therapist then instructs the woman to begin to go to places that remind her of the attack. Over time, this intentional exposure to cues and memories of the trauma shifts the so-called "locus of control" to the victim, who realizes that she can control her unpleasant and intrusive thoughts.

Foa, who is an Israeli, has taught her technique to therapists with the Israel Defense Forces. These therapists recently treated thirty soldiers who had severe PTSD. Some had been in continuous psychotherapy until they received Foa's treatment, which typically requires only twenty hours of therapy. Twenty-nine of the thirty experienced a marked improvement in both their symptoms and their ability to function.

Neuroscientists and experimental psychologists are now mapping the circuits in the brain that could account for the success of Foa's treatment. For example, rats exposed to a tone and then given an electric shock learn to associate the tone with the shock, so that simply hearing the noise causes them to exhibit increased pulse, muscle contraction, and avoidance behavior — an analogue to PTSD. If the tone occurs without the shock being given and is repeated on multiple occasions, the rats no longer respond with these anxiety symptoms. In a related experiment, Joseph LeDoux, a neuroscientist at New York University, made lesions in the prefrontal lobes of such fear-conditioned rats — in a part of the brain just behind the forehead. He then provided the tone without administering the shock; the animals were unable to extinguish their anxiety response, which suggests that the missing circuits play a critical role in stress management.

In recent years, Foa's technique has been used not only to treat PTSD but also to prevent it. Richard Bryant, a psychologist in Australia, has treated people who displayed sustained symptoms of acute anxiety after a motor vehicle accident or an assault. In three randomized controlled trials, six months after the trauma, patients who had received treatment were three times less likely to develop PTSD compared with members of the control group, which received only supportive counseling.

Despite considerable evidence in the United States and abroad showing that treatments like those developed by Foa can ameliorate established PTSD — and possibly help prevent the disorder in people with acute stress reactions — her approach has not been widely adopted. Most counselors find cognitive-behavioral techniques unappealing. Dr. Steven Hyman is a neuropsychiatrist and the provost of Harvard University; in 2001, he was the head of the National Institutes of Mental Health. "When I was NIMH director, I was upset by how few people wanted to learn cognitive-behavioral therapy," Hyman told me. "Here was a therapy proven to be effective by clinical trials. But psychologists and psychiatrists are so interested in people, and they want to cure you with their understanding and empathy and connection. The cognitive-behavioral approach is by the book, mechanical, pragmatic. The therapists find it boring. It's not their idea of therapy, and they don't want to do it." Debriefing holds more allure for most counselors, for it reflects a prevailing cultural bias; namely, that a single outpouring of emotion — one good cry — can heal a scarred psyche.

Foa's method has begun to find some adherents. Malachy Corrigan, of the FDNY, now uses cognitive-behavioral techniques with several groups, including firefighters who narrowly survived the collapse of the towers. In November 2001, Foa came to New York and trained forty therapists in her technique. Now Columbia University is offering seminars to therapists who are interested in learning Foa's approach.

At the same time, the scientific critique of debriefing has begun to have an impact. The Department of Defense, the Department of Justice, the Department of Veterans Affairs, the American Red Cross, and the Department of Health and Human Services have all abandoned it as a therapeutic method. Bruce Blythe's company, Crisis Management International, which is based in Atlanta, recently decided to discontinue its debriefing service. This week, the American College of Neuropsychopharmacology Task Force on Terrorism will release a paper recommending that debriefing be abandoned as a mainstream prevention method. Nevertheless, many for-profit companies in the so-called "grief industry" continue to offer single counseling sessions that are fundamentally linked to Mitchell's seven-step technique. And debriefing is still widely embraced; counselors for the NYPD and the Los Angeles Fire Department continue to use the method.

Perhaps the solution, Hyman said, is to drop the idea that "counseling" is necessary. He told me that the way we respond to individual or mass trauma should be guided by how we behave after the loss of a loved one. "What happens when someone in your family dies?" he said. "People make sure you take care of yourself, get enough sleep, don't drink too much, have food." Hyman pointed out the different rituals that various cultures have developed — shiva among Jews, for instance, and wakes among Catholics — which successfully support people through grief. "No one should have to tell anyone anything!" he said. "Particularly not in the scripted way of a debriefing." The traumatized person should share what he wants with people he knows well: close friends, relatives, familiar clergy. "It's so commonsensical," Hyman said. "But the power of our social networks — they are what help people create a sense of meaning and safety in their lives."

JOHN HORGAN

Keeping the Faith in My Doubt

FROM *The New York Times*

WITH the presidential election over and the holidays upon us — a religiously charged political season followed fast by the most religious time of the year in an overwhelmingly religious nation — unbelievers may be feeling a bit beleaguered. To cheer themselves up, they might visit the virtual home for a group called the United Universists.

Founded last year by a few brave souls in Birmingham, Alabama, the Universism movement "denies the validity of revelation, faith, and dogma" and upholds science as our most reliable source of truth. The Universists are asking atheists, agnostics, and other infidels to join them in their effort to counter the influence of religious zealots in our culture. Since the recent election, the Universists have posed this question on their home page in large type: "Who will fight for the faithless?"

Good question. Obviously neither major political party wants to associate itself too closely with unbelievers — and understandably so, given polls showing that Americans are even less likely to vote for an atheist for president than for a homosexual. But as an areligious person myself, I'm intrigued by the notion of unbelievers banding together to increase their political clout, perhaps by speaking out on issues like sexual freedom, abortion, stem cell and cloning research, and prayer in schools.

There are more of us heathens out there than you might guess. According to the Pluralism Project at Harvard, which tracks religious diversity in the United States, the number of people with no religious affiliation has grown sharply over the past decade, to as many as 39 million. That is about twice the number of Muslims, Jews, Buddhists, Hindus, and Episcopalians combined.

Not surprisingly, a slew of organizations — including older ones like the Council for Secular Humanism and the American Atheists and newer ones like the Universists and the so-called Brights — are competing for the devotion of the godless. The Universists, who claim to have enlisted 5,000 members so far, are especially feisty and shrewd at self-promotion. In September they took to the streets of Birmingham to protest Alabama's ban on the sale of sex toys, and last week they organized an on-line chat with Sam Harris, author of the antireligion polemic *The End of Faith*.

And yet I have no plans to sign up with the Universists or any other areligious group. First of all, I'm just not a joiner, more out of laziness than anything else; I avoid commitments that might jeopardize my sports- or sitcom-watching time. An organization for freethinkers — one of the Universists' self-definitions — also strikes me as oxymoronic, like an anarchist government. Isn't the point of being a freethinker eschewing categories like Satanist, Scientologist, or Universist?

I'm also disturbed that these areligious groups have exhibited the same sectarian squabbling that they deplore in religious believers. When Michael Shermer, publisher of *Skeptic* magazine and director of the Skeptics Society, was invited to speak at an atheism convention in Florida last year, some organizers objected because he is agnostic — a mere doubter of God's existence rather than a denier. Shermer has likened this hairsplitting to the dispute between Baptists and Anabaptists over whether baptism should take place during infancy or adulthood.

At that same conference, two antireligion educators also proposed that negative terms like "agnostic," "atheist," "unbeliever," and "skeptic" be replaced with the more upbeat "bright," which describes someone "whose worldview is naturalistic — free of supernatural and mystical elements." The term, which can serve as a noun or an adjective, has been promoted by the philosopher Daniel Dennett and the biologist Richard Dawkins.

Members of some other groups have reacted with annoyance to the Bright movement, no doubt seeing it as an intrusion on their turf. Defenders of the old standbys "atheist," "agnostic," and "secular humanist" complain that "bright" is self-aggrandizing —

and the implied antonym, "dim," a tad demeaning. Critics of the Brights include the Universists, whose Web site also distinguishes Universism from (and not so subtly asserts its superiority to) atheism, deism, humanism, pantheism, transcendentalism, and Unitarian Universalism.

All this goes to show that even groups founded with the best of intentions — and what groups aren't? — usually become concerned above all with self-perpetuation, often at the expense of other groups with similar aims.

My main objection to all these antireligion, pro-science groups is that they aren't addressing our basic problem, which is ideological self-righteousness of any kind. Obviously, not all faithful folk are intolerant bullies seeking to impose their views on others. Moreover, rejection of religion and adherence to a supposedly scientific worldview do not necessarily represent our route to salvation. We should never forget that two of the most vicious regimes in history, Nazi Germany and the Soviet Union under Stalin, were inspired by pseudoscientific ideologies, eugenics and Marxism.

Opposing self-righteousness is easier said than done. How do you denounce dogmatism in others without succumbing to it yourself? No one embodied this pitfall more than the philosopher Karl Popper, who railed against certainty in science, philosophy, religion, and politics and yet was notoriously dogmatic. I once asked Popper, who called his stance critical rationalism, about charges that he would not brook criticism of his ideas in his classroom. He replied indignantly that he welcomed students' criticism; only if they persisted after he pointed out their errors would he banish them from class.

Of course we all feel validated when others see the world as we do. But we should resist the need to insist or even imply that our views — or antiviews — are better than all others. In fact, we should all be more modest in how we talk about our faith or lack thereof.

For me, that isn't difficult, because I've never really viewed my doubt as an asset. Quite the contrary. I often envy religious friends, because I see how their faith comforts them. Sometimes I think of my skepticism as a disorder, like being colorblind or tone-deaf. Perhaps I'm missing what one geneticist has called "the God gene," an innate predilection for faith (although I'm skeptical of that theory, too). But skepticism has its pleasures; I like the feeling of traveling lightly through life, unencumbered by beliefs.

Instead of banding together, maybe we unbelievers should set an example by going in the opposite direction. We should renounce all "isms" that claim to speak for our most profound personal beliefs. Or rather, since we seem to be headed in this direction anyway, each unbeliever could create his or her personal ism, perhaps with its own name. Since Universism is taken, I'll call mine "Horganism." You can revile it, admire it, or ignore it, but you can't join it.

JENNIFER KAHN

The Homeless Hacker v. The New York Times

FROM *Wired*

NOT LONG AGO Adrian Lamo was exploring an abandoned gypsum processing plant in West Philadelphia with two friends, when a police cruiser drove slowly by. Lamo's friends were high on methamphetamines, and at the sight of the cops they urged him to run. Instead, Lamo stood still, and as he did, he heard a strange rasping sound. Peering down a nearby sewer grate, Lamo found the source: a kitten, meowed to hoarseness, scrambling around on a pile of trash.

When a second squad car pulled up and fixed its spotlight on Lamo, he walked forward. "I said, 'Officer, I'm so glad you're here!'" Lamo recalls, using his most innocent voice. "'There's a kitten trapped down here.'" The officer was suspicious, but two hours later, with the assistance of three additional cruisers and a police van, the kitten was out.

"There we were," Lamo says. "Circle of police cars, flashing lights, me — quasi-notorious cybercriminal — and my coconspirators all working in concert to try and get this kitten out of the drain." In the hubbub, neither Lamo nor his friends got searched, and the kitten went home with them in a discarded juice carton. Lamo named it Alibi.

Being saved against all odds is a theme with Lamo, who told me the story of Alibi shortly after he was arrested by the FBI in September for computer fraud. The charges — that Lamo had broken into the private network of the New York Times Company and run

up a $300,000 bill on the pay-per-use search tool Lexis-Nexis —
carried a possible fifteen-year prison sentence. Asked if he was
afraid of going to jail, Lamo said simply, "I'm sure it would be educational. The beautiful thing about the universe is that nothing
goes to waste."

Thin and pale, Lamo has a delicate, androgynous face and a
habit of hunching his shoulders as though to stay warm. He is one
of the best-known hackers in the country and was out being filmed
for a documentary when the cops came looking for him at his parents' house in Sacramento. While cameras rolled, Lamo described
his most famous hacks, a string of highly publicized computer
intrusions — Microsoft, AOL, and Excite@Home — of which the
Times was merely the most recent. Just months before the *Times*
hack, he had made the papers by burrowing into WorldCom's
intranet, where he found a database containing Social Security
numbers, bank account data, and direct deposit instructions for
some 86,000 WorldCom employees — plus a Web router maintenance tool that enabled him to go deep into the private networks
of Bank of America, Citicorp, and JP Morgan.

Known as the Homeless Hacker before his arrest, Lamo did most
of his virtual exploring from the Internet connections at Kinko's
copy shops. Besides his laptop — an eight-year-old Toshiba with six
keys missing — he traveled light, usually with a blanket, a change
of clothes, and a Taser stun gun, which he used to pick electronic
locks and sometimes to shock vending machines to see if they
would drop food or spare change.

Relentlessly nomadic — he has crossed the country by bus half a
dozen times — he's also a connoisseur of serendipity. He once
spent two weeks attending a Pennsylvania Bible school on a whim.
Mostly, though, he transited between Washington, D.C., and San
Francisco, where he grew up. Because he has friends in those cities,
Lamo could usually count on finding a place to sleep that was both
more secure and more pleasant than his usual home, an abandoned building. The capital also has the virtue of being an information hunter's paradise. "In D.C., it's hard to open a dumpster
without finding classified documents," he tells me wistfully.

In return for food and a place to sleep, Lamo took his hosts on
rambling adventures: through city sewer systems or locked office
buildings. The tours were always surreal, one friend recalls: "Not so
much fun as different."

"To me," Lamo explains, "ending up in a city that I've never been to before, with no money, where I know nobody, and yet somehow making it work out is as much a unique and amazing exercise of faith as going to a computer network that I know nothing about and somehow finding myself in its innermost recesses."

Lamo prides himself on his ability to get out of tight situations, and once he became a wanted felon, he briefly dropped out of sight. But the FBI, prompted by fears of Internet terrorism, was only getting started. In the end, even the Justice Department got involved, unsuccessfully attempting to subpoena the notes of a dozen reporters who had interviewed Lamo. The game had suddenly turned serious. After five days as a fugitive, Lamo surrendered in a Sacramento Starbucks.

For someone with such a glamorous hacking resume, Lamo is strangely unschooled. Illiterate in computer languages like Java and C++, he cannot exploit loopholes in a system's underlying code. Instead, he uses a common man's tool: the Web browser. Firing up Internet Explorer, Lamo will troll through a corporation's homepage, seeking outsourced jobs like advertising, distribution, and payroll. The companies that handle these tasks link to the main corporate database, but their proxy servers — the point of connection between the two networks — are often poorly secured, sometimes with standard-issue passwords that no one bothers to reset.

Finding these weak points is a matter of perseverance more than talent, but Lamo has also been unusually lucky — often materializing in areas of corporate networks that were heavily guarded and distinctly off-limits. One fellow hacker describes the skill — with a nod to the sci-fi author Neal Stephenson — as "the ability to condense fact from the vapor of nuance."

It was this vaporous instinct that in February 2002 first drew Lamo to the *New York Times* servers. Messing with the *Times*'s news site would be a coup, Lamo knew, but the Gray Lady had been hacked this way once before, and security was tight. Rebuffed from the news server, Lamo focused on the corporate network, sending test e-mails to the paper's autoresponder, culling IP addresses, and finally stumbling onto a subnetwork that controlled, among other things, the database containing information about op-ed writers. This being the *New York Times,* the list of contributors was particularly luminous, an establishment who's who that included UN weapons inspector Richard Butler and former National Security Agency head Bobby Inman, as well as celebrities like Robert Redford and Rush Limbaugh. Many of the names had phone numbers and home addresses attached, along with notes on the person's area of expertise, payment history, and editorial temperament. After browsing a bit, Lamo added himself to the roster, brazenly giving his full name and cell phone number. (For his expertise, he dryly listed "computer hacking, national security, communications intelligence.") Once the deed was done, he called SecurityFocus .com reporter Kevin Poulsen, a confidant and a convicted hacker himself. Lamo gave him the scoop, and Poulsen, hoping to verify the story, called the *Times*.

Publicly exposing the company whose system he has just compromised is Lamo's MO, and for the most part it has served him well. WorldCom, for instance, officially thanked Lamo after he ignored the temptation to steal a quick million in paychecks; he later spent a weekend briefing managers on the details of his work. Excite@Home also offered thanks. The *Times* did not feel similarly indebted. Notified of the breach, the company alerted the U.S. Attorney's office, which began an investigation in February 2002. By shadowing the twenty-three-year-old hacker, the agencies learned that Lamo had done more than just add his name to the list of op-ed writers. He had also created several passwords that gave him access to the paper's account with Lexis-Nexis and its massive database of public records and thousands of newspapers and magazines.

Lamo, it turns out, spent several months playing on Lexis-Nexis, mostly digging for personal information on other hackers and on journalists who had written critically about his work. At one point, he also attempted to find all the license plate numbers assigned to undercover vehicles registered with the FBI. All told, he conducted more than three thousand searches. Although the *Times* doesn't pay retail for the service, the FBI calculated Lamo's damages using the full Lexis-Nexis rate, which added up to a shocking $300,000. It was clearly a punitive figure. Had Lamo simply bought an unlimited three-month account with Lexis-Nexis rather than piggybacking off the *Times*, it would have cost him just $1,500. But for the *Times*, it was not the money so much as the principle: "A serious offense," says a *Times* spokesperson. No one was particularly grateful to Lamo for pointing out the vulnerability in the op-ed database, and some even saw the action as sinister: a subtle attempt to divert attention from the real ongoing theft. When the FBI concluded its investigation last summer, the company decided to press charges.

When it comes to ethics, hackers fall into three main categories. The good guys — the whitehats — have jobs with computer security firms and work strictly within the law. The blackhats break into networks illegally, usually to steal or vandalize. Lurking in the middle are the grayhats — hackers who are not openly destructive but who get their thrills from joyriding through private systems or conducting uninvited "security checks." In a typical example, a pair of hackers known as the Deceptive Duo defaced dozens of military and commercial Web sites, plastering their homepages with a picture of the American flag, crossed pistols, and a warning to "Tighten security before a foreign attack forces you to." Grayhats see themselves as Internet Zorros — high-minded vigilantes who are righteously setting information free while nobly helping to protect it from vandals. In practice, though, it can be hard to tell the noble outlaw from the petty criminal. Breaking the law in the name of improving the law is rarely condoned, let alone idealized. The line between self-interest and "setting information free," moreover, is easily blurred — and it's the murky middle ground in the already ill-defined grayhat arena where Lamo most likes to operate.

Exactly what shade of gray hat Lamo wears remains a matter of heated debate. On Slashdot, a Web site popular with the computer security crowd, members spent days bickering over the most appropriate metaphor for Lamo's corporate hacking. Was it a purely good deed, like walking by an unlocked car full of money and alerting the owner? Or was it much creepier, like rattling the locks on a neighborhood house, then leaving a note on the bed telling the owner that she left her bathroom window open?

In theory, it's easy to see Lamo as a good guy. Unlike many hackers — even whitehats — he never uses a pseudonym and makes no effort to hide his identity. If the company he notifies appears grateful, he will often offer to help plug the hole he's discovered for free. Poulsen, for one, believes that Lamo "practices a style of hacking — open, brash, illegal, but carefully observant of an unwritten code of ethics — that went out of style a decade ago."

Indeed, Lamo's hacks are uncommonly witty and at times almost inspiring. Once, after tunneling into Excite@Home's customer service database, Lamo pulled the email and phone number of a customer whose complaint had gone unanswered for a year. Lamo called him up, chatted briefly, then offered to forward him all the company's internal correspondence pertaining to the original complaint.

Stunts like this have made Lamo a legend in hacker circles. His Friendster account is full of admiring testimonials from fellow geeks. Craig Calef, a twenty-two-year-old engineer from Massachusetts, writes, "I'm not a religious man, but Adrian is the closest thing I have to a messiah." Nevin Williams, who was a lead operations engineer for Excite@Home at the time of Lamo's hack, says, "It's unclear whether you should be amused, comforted, concerned, inspired, or indicted for the privilege of not ever being bored in his presence. He is the Man's Man, though the Man doesn't seem to know he's been hacked yet."

At his best, Lamo seemed to be operating not just outside the law but above it. And though supporters praise this kind of freelance justice, detractors question Lamo's motives. Given his habit of alerting the media to every hack, many on both sides believe that Lamo is driven largely by vanity. Fellow hacker Mike Sanders maintains that Lamo didn't actually discover the Excite@Home security hole — he just made it public and took the credit. Lamo has also demonstrated his techniques for MSNBC, and the outgoing message on his cell phone (666-HACK) leads to a line offering detailed instructions for reporters on deadline.

Moreover, Lamo's unwritten code of ethics includes some disturbing fine print. If he feels that a store clerk has been surly, for instance, he's not above picking through dumpsters for that employee's personal information. Three years ago, when Lamo had a falling-out with several contributors on Observers.net, an anti-AOL site, he took revenge by appropriating the offenders' on-line identities. His ex-girlfriend, who requested that her name be withheld, has also accused him of stalking her. "Every time I moved, he'd send an anonymous e-mail," she remembers. "Sometimes he'd include my unlisted phone number, which nobody else had. He made a point of showing that he knew where I was." The court issued a restraining order against Lamo, following a complaint in which his then girlfriend described an ongoing pattern of harassment and abuse. "He carried a stun gun, which he used on me," she recalls. "He was very controlling. He wanted to know where I was constantly. After we stopped talking, he would hack into the phone company's service center and change my phone services."

Since his arrest in September, Lamo has been living with his parents in Carmichael, a suburb of Sacramento. His computer use is restricted, and he must check in with his pretrial services officer regularly — ostensibly to prove that he hasn't skipped town. In practice, however, this system seems rather lax. For one recent check-in, Lamo used a voice-over-IP connection that gives outgoing calls an out-of-state number and area code. The officer didn't blink, Lamo reports. "If he'd checked his caller ID, it would be like: 'Gee, this call is coming from Connecticut!'"

For our first interview, Lamo suggests that we meet at a Starbucks across town from the one where he gave himself up six weeks before. His parents, he explains, are "poorly disposed" toward reporters. This is disappointing, since I've been curious about his parents, who strike me as almost unfathomably tolerant. According to Lamo, in order to post the $250,000 bail that resulted from his arrest, the Lamos — who also have a five-year-old daughter and an eleven-year-old son — had to put a lien on their house. And yet so far they have not insisted that Lamo get a job or even that he stop breaking into corporate computer systems. "They're dream parents," says Darci Wood, who helped found FreeLamo.com, a Web site dedicated to raising money for Lamo's legal expenses. "It's amazing the way they've supported him in the face of these charges." (Wood is the girlfriend of the notorious Kevin Mitnick, who spent five years in jail for hacking his way into the phone system.) When I ask Lamo what his parents think about his ongoing career in computer trespass, he says with practiced wryness, "They think it's about time for me to do something that doesn't involve mandatory federal sentences."

In person, Lamo is not quite as I imagined him. He's beset by facial tics, including one that makes it look like he's winking. He is, however, disarmingly polite, offering me half of his pastry and apologizing when the din from a Frappuccino blender threatens to overwhelm my tape recorder. Candid to the point of incrimination, he also cheerfully recounts the details of "my alleged Lexis-Nexis foray," and lets me in on several new intrusions. Among other things, he says, he now has the ability to shut off the FBI's phone service. "All of it?" I ask incredulously. "Well, the field offices," he demurs. "There's a lot of FBI."

Lamo began hacking into phone systems in high school. It was a lonely time, and he eventually dropped out with an equivalency degree. After a short stint doing computer work for PlanetOut.com, Lamo began to roam. At one point, he spent a month stranded in a small California town outside Visalia, too poor to afford Greyhound and unwilling to accept his parents' offer of forty dollars to cover the ticket. (His mother compromised by mailing him soup.)

Perpetually broke and often hungry, Lamo would also rely heavily on the material advantages of his more solvent friends. (While in Visalia, he slept on an acquaintance's floor for a month.) Nevin Williams, the former Excite@Home engineer, regularly offered Lamo a place to stay when he was in San Francisco but now says he feels manipulated. "If there's a generous person, and you make them aware of a need in a roundabout manner, the generous person might be inclined to offer help," he says dryly. "I think Adrian understands this very well."

Lamo is an eager interview subject, but the longer we talk, the more anxious he gets. At one point he stops midsentence and tips his head meaningfully toward a man stirring his coffee at the counter next to us. An FBI eavesdropper? I say that maybe we should go. "It's OK," Lamo shrugs. "They probably have the Starbucks bugged anyhow." He pulls on fleece gloves and slides his hands along the underside of the table. "I used to carry a frequency counter," he says. "Now I can't afford one."

As we're leaving, Lamo points out another customer, who left when we got up and is now sitting in his car reading the paper. Lamo has me drive us around the parking lot a few times, then jumps out and raps on the guy's car window. The man rolls it down.

"Do you now or have you ever worked for the government?" Lamo asks. The man looks startled and wants to know what's going on. "Oh, I'm an accused felon," Lamo volunteers cheerfully. "Alleged to have compromised certain computer systems." Back in the car, he giddily shows me the man's business card. He's happy again, with the residual charge of someone who's just survived a close shave. I, however, am suddenly peevish. "Wouldn't any self-respecting undercover FBI agent carry a fake business card?" I ask. Lamo's face falls. We leave the parking lot without further discussion.

Lamo's paranoia is infectious. Not long after our first meeting, I interview two of his friends, both of whom joke that he has probably tapped my phone. This seems improbable, but after a while, I can't seem to shake the idea. A few days later, Lamo calls at midnight. He patches me through to an IRS recording at the point where they're asking for a Social Security number, tells me to enter mine, and promises he won't listen. When I refuse, he gets angry. "I'm doing this for you," he huffs. But when I ask what, precisely, he is trying to do, he says that he won't discuss it over the phone.

In another late-night phone call, Lamo reveals that he recently went to a doctor, again announced that he was "an accused felon," and said that he wanted his life to be less stressful. To Lamo's annoyance, the doctor gave him prescription sleeping pills and a four-week supply of Paxil, which he refuses to take. He was hoping to score Xanax, which he says is good for short-term anxiety, panic attacks, and insomnia. "If you say 'I want Xanax' up front, they say 'Oh boy, only Paxil for you!'" Lamo explains. "But if I go back in two weeks and say, 'The Paxil isn't working, plus I've started getting these little, like, electric shocks throughout my body' — then they think you're showing signs of petit mal seizure, which is one of the side effects, and they give you Xanax."

Lamo actually is seizure-prone, ever since he overdosed on prescription amphetamines back in 2001. Now, he explains, he sticks to depressives and dissociatives. "The dissociatives are amazing," he boasts. "You can look at your face in the mirror and completely not recognize it."

Not long after this exchange, Lamo leaves a message that he's "concerned about the direction" that some of my reporting has taken. He also calls my editor, informs him that I have been asking "inappropriate" questions about his mental health, and threatens to stop cooperating with the article. It's an uncharacteristic reaction for Lamo, who in the past had treated the idea of being profiled with unnatural enthusiasm — at one point even reassuring me that a negative portrayal of him would be fine. What pushed Lamo over the edge, I learned, was that I had asked his friends about his drug use and whether they ever worried about his health. (For what it's worth, most said that they do worry, but that Lamo can take care of himself.) Lamo eventually rescinded his threat, although the fact that he made it at all is revealing.

Getting hacked is a creepy feeling — one that has more to do with privacy than actual damage. After all, the real reason the *New*

York Times called the FBI is not because Lamo mooched off its Lexis-Nexis account but because he dug through a database that was confidential and personal. Yet the anxiety of exposure cuts both ways. Ironically, the very thing that made Lamo so uncomfortable is precisely what he does to the rest of the world. Once I began sifting through Lamo's personal information — even with an invitation — he reacted much as the *Times* did. Rather than trusting his friends to be kind, or me to be fair, he panicked and called the authorities.

In January, nearly two weeks after I last spoke to Lamo, news comes down that he has agreed to a plea bargain. By confessing to a felony charge, he will receive a shortened sentence of six to twelve months and pay no more than $70,000 in damages. Copping to a felony that will plunge him into debt doesn't seem like much of a victory, but when I call Lamo afterward, he sounds surprisingly upbeat. "I always said that I would accept the consequences of my actions," he says. While acknowledging that the past week had been difficult, he is quick to move on to happier matters. Just recently, he reports, he was offered the chance to write a monthly security column at $500 a shot for a new magazine on mobile computing. He has talked before about hanging up his spurs to become a journalist, and this should be a dream job for an aspiring technology reporter. Lamo, however, has already dismissed the offer. The magazine will "popularize" his copy, he insists. "For me, it would have to be either exactly as I put it down on paper or not all."

This seems a bit high-handed to me, and I tell Lamo that I think he's being foolish. Now that he's saddled with a felony conviction and thousands in debt, how choosy can he be? Lamo, however, will not be swayed. "I think the editors see the column the way McDonald's sees flipping burgers, while I see it as an effort of religion," he muses. As to his future, Lamo says merely that he hopes the universe will provide. What he doesn't realize is that the universe has provided, just not in the way that he wants. After all, the chance to be a columnist is nearly as miraculous as finding a kitten in a gypsum factory. But to reinvent himself, Lamo would have to shed his life as the Homeless Hacker — and that's something he seems unwilling to do. Even after he has served his time, Lamo will remain a prisoner of his own myth, picking through the rubble of life on the margins.

ROBERT KUNZIG

20,000 Microbes Under the Sea

FROM *Discover*

THE THING ABOUT THE MUD hoisted from the bottom of the Black Sea in the summer of 2001 — the thing that surprised and delighted the researchers aboard the *Professor Logachev* — was that there was hardly any mud at all. They were 75 miles west of the Crimean city of Sevastopol, 750 feet above an undersea slope along which sediment from the Dnieper River cascades down into the depths. With a set of giant steel claws guided by a video camera, they had taken a one-ton bite out of that slope and dumped the goopy mess on deck. It stank. That didn't surprise anyone — sea floor mud often contains hydrogen sulfide, which smells like rotten eggs. But they were struck by what else was in the sample: "Nearly one ton of biological material," says Walter Michaelis, a biogeochemist at the University of Hamburg in Germany, who led the expedition. "No sediment. No carbonates. It was a cubic meter of bacteria!"

A few days later, Hamburg geologist Richard Seifert got a look at the sea floor where the bacteria came from. Diving in a little German submarine called *Jago*, peering through an acrylic porthole several inches thick, he saw a fog of floating particles. Then Seifert saw, looming out of the dark and into the sub's tight halo of floodlight, through curtains of rising bubbles that made it seem as if the sub were driving through champagne, chimneys — black, knobby spires, the tallest rising more than thirteen feet off the sea floor. Chimneys made by erupting volcanic minerals are common at hot springs in midocean ridges. But this was not a hot spring, and these chimneys were not built of volcanic rock. The pilot prodded one with *Jago*'s hydraulic arm. It was soft, like flesh. He knocked one over, felling it as if it were a tree to reveal its cross section. Under a black outer layer there was a thick layer of pink and a core that was harder and greenish gray.

The chimney had been made entirely by single-celled microbes. The microbes formed the outer layers; the hard core was a carbonate mineral they had secreted. What Seifert and his colleagues had discovered was an outcrop — of life. The evidence is now clear that far below the sea, and far below the floor of the sea, in sediments all over the world, microbes live to astonishing depths — the record so far is half a mile — and in astonishing numbers. The deepest of the microbes make methane, which the ones in shallower sediments consume. To all of them, oxygen is poison. They are relics of an early period in Earth's history when methane was abundant and green plants had not yet given the planet oxygen. "Maybe the early Earth was all covered with blackish, pinkish slime," says Antje Boetius, a biogeochemist and geomicrobiologist at International University in Bremen, Germany.

Today these ancient organisms have been pushed into obscure niches where oxygen does not penetrate. Mostly that means below the sea floor — except in a few special places like the Black Sea. With only the narrow Bosporus as an outlet, the Black Sea is seldom flushed, and the oxygen in its deep water, below 600 feet or so, has long since been depleted. No fish live at that depth. But microbes do, thriving on methane that bubbles up from below. The methane is made by the microbes' deeper cousins, and they are numerous.

The total mass of microbes living beneath the sea floor has been estimated at as much as a third of all the living stuff on the planet. The total amount of methane made by these microbes is probably greater than the mass of all known reserves of coal, gas, and oil. Methane is a potent greenhouse gas, and huge belches of microbial methane from deep reservoirs, where it resides mostly as frozen methane hydrate, have been linked to rapid changes in Earth's climate. They may have helped pull the planet out of recent ice ages, and they almost certainly helped end the Paleocene Epoch 55 million years ago with an intense burst of global warming. Nor are the potential impacts of deep-sea methane limited to climate. Blasts of it have been linked, in respectable journals, to mass extinctions, to undersea landslides that caused ocean-crossing tsunamis, and even to the mysterious disappearance of ships at sea.

All that may seem a lot of action to attribute to mud. But that's precisely the essence of what researchers have been finding lately: sea-floor mud is alive, and it is powerful. It's the whale we managed not to notice until now.

The thing about the mud that marine geologist and geochemist Erwin Suess hoisted from the bottom of the Pacific in 1996, half a mile down and 60 miles off Newport, Oregon, was that it was seething — cold, but seething. On the TV monitor aboard the research vessel *Sonne,* he could see the mud as it was brought to the surface. A few minutes later, on deck, "the whole cubic meter of stuff was bubbling," he says. "It was just one smelly olive green mess of sediment. People were hesitating, standing back, because it was just an awful smell, the hydrogen sulfide. So I rolled up my sleeves and reached into that mess and stood up to my ankles in it, and then touched this hard stuff — it was cold, ice cold. I threw it on the deck to get rid of the sediment, because it was all covered, and it broke open. It was pure white."

A lot of photos were snapped on the *Sonne* that day and on later cruises to the same spot. The photos show rubber-gloved hands holding slimy hunks with the color contrast of a coconut: brownish on the outside, white on the inside. In some of the pictures you can see little craters in the white stuff, the hemispheric outlines of vanished bubbles. In others you see people boring into it with power drills, slicing it with knives, plunging it for safekeeping into basins of liquid nitrogen. You see young scientists grinning ear to ear as they hold pieces of the white stuff in their hands — showing the camera how it can be ignited, how it burns brightly with an orange flame.

The white stuff is methane hydrate, and it is weird. Researchers call it icelike. It is frozen solid, but it is not exactly ice. It is a form of clathrate, a term derived from the Latin word for cage. Methane hydrates consist of molecules of methane trapped in cages of H_2O. Typically, six water molecules surround a single methane molecule, and many of those cages linked together form a crystal. Only at high pressures can methane insinuate itself into water this way. But if the pressure is more than thirty times the normal atmospheric pressure — easily exceeded under a thousand feet or more of water — hydrates can form at temperatures above 32 degrees Fahrenheit. This usually happens just beyond the continental shelf, where it slopes down into the abyss.

There is a precise curve of temperatures and pressures that define the depth at which methane hydrates exist. The bottom of that zone, deep in the sea-floor mud, is where the temperature gets too high, toward Earth's hot interior; the top of the zone is where the pressure gets too low, moving toward the surface. Leave the zone in either direction and the partnership of methane and water dissolves. When you bring a chunk of methane hydrate up to the surface, the water melts and drips through your fingers; the methane gas wafts into the air.

Hydrates have been a curiosity for nearly two hundred years. Natural methane hydrates were first discovered by Russian scientists in the late 1960s in Siberian permafrost — where the ground is so cold that hydrates can form at shallower depths and at lower pressures than under the sea — and then, in the 1970s, at the bottom of the Black Sea. (So much methane comes out of the Black Sea that sailors have reported seeing lightning ignite it at the surface.) It was not until the 1980s that researchers drilling into the ocean floor first began to understand that this stuff is everywhere. Usually, though, the frozen hydrates they brought up in cores were small, like pencil erasers.

The chunk that Suess brought up off Oregon was a foot and a half across. There was plenty of it to study. Over the years, as researchers have returned to the place they named Hydrate Ridge, they have learned a lot about how methane hydrate is created there. Hydrate Ridge is one of several accretionary ridges off Oregon, long layers of mud that get scraped off the Pacific tectonic plate as it plunges under the North American continent. Methane gets squeezed out of the deepest layers of sediments like water from a sponge and migrates up toward the sea floor. The southern summit of Hydrate Ridge, about 2,500 feet below the sea surface, is a field of mounds and depressions 10 to 20 feet across. It was from one of those mounds that Suess's team grabbed chunks of thick, pure methane hydrate.

There is a lot of methane under Hydrate Ridge. Under most frozen hydrate deposits is a layer of free methane gas occupying the pore spaces in the sediment. Typically, that's how the deposits are discovered, because the boundary reflects sound well enough for a survey ship to detect it. Under some parts of Hydrate Ridge there is so much methane gas, says German geologist Gerhard Bohrman, that it is constantly bubbling up into the hydrate zone. There is so much methane that, as it freezes instantaneously to form hydrate, it draws all the water out of the sea-floor ooze and dries it out completely — and often there is methane left over, trapped as large bubbles in the porous hydrate. Bohrman proved that by bringing a sediment core up in an autoclave. He kept it under pressure while he had it CT-scanned in a clinic in Palo Alto. Before that he and his shipmates had seen how buoyant the hydrate was: as they worked off the coast, large blocks of it sometimes bobbed to the surface near the ship.

There is so much methane rising up under the southern summit of Hydrate Ridge that some of it bubbles all the way to the sea floor. "You build up too much free gas, and then you have an over-pressured column," says Gerald Dickens, a marine geochemist at Rice University who went to Hydrate Ridge on a drill ship in 2002. "And the gas just cracks the sediment and migrates right up to the sea floor." Sea floor gas chimneys have been turning up in many places, Dickens says, now that researchers know how to recognize them on seismic readouts. And places where methane bubbles into the seawater, as it does at Hydrate Ridge, are commonplace around the world.

Ultimately, that methane is derived from bigger and more complex organic compounds in the buried sediment. That's why hydrates, like oil — and like fish — tend to be found along the world's coastlines, where the waters are rich in nutrients and plankton corpses fall like thick snow to the sea floor. It used to be thought that the methane in hydrates was made the way oil is — that Earth's internal heat made methane, the smallest hydrocarbon, by cracking bigger hydrocarbons at depths of more than a mile below the sea floor. But then researchers started looking at the carbon isotopes in hydrates. They found that most hydrates, compared with the sediments around them, are enriched in the isotope carbon-12 and depleted of the heavier carbon-13. Heat would not be choosy that way about the molecules it cracks. Life, however, is choosy: all living things selectively take up carbon-12 and reject carbon-13.

The carbon-isotope ratio of sea-floor hydrates indicates that the methane was made by microbes. "These microbes are forming enormous amounts of gas," says Dickens. "But it's not like the hydrates are just building up over time, because we're also losing methane out of these systems." The puzzling thing is that methane isn't bubbling up *everywhere*. But that puzzle was solved at Hydrate Ridge.

Antje Boetius was a young biogeochemist aboard one of the research expeditions off Oregon in 1999. She had recently moved to the Max Planck Institute for Marine Microbiology in Bremen, which collaborates closely with Suess's group, to learn a couple of fundamental techniques of molecular ecology she needed to complete her own research project in the Indian Ocean. That project was not related to hydrates or to the deep biosphere. It was "just regular deep-sea research — going to the big wide cold ocean," Boetius says now, as one who has left all that behind. The discovery she made at Hydrate Ridge changed her career.

What captured Boetius's imagination there were the clusters of organisms, known as cold-seep communities, that had taken up residence around the places where methane seeps from the sea floor. Suess was among the first researchers to discover a cold-seep community. While he was working at Oregon State two decades ago, some of his colleagues explored a sea-floor hot spring near the Galápagos Islands and brought back specimens of giant white clams collected with the submersible *Alvin*. The first time Suess got to dive in *Alvin*, in 1984, it was nearly 5,000 miles northwest of the Galápagos, just a few miles west of Hydrate Ridge and nowhere near a hot spring. Yet "there were the same damned critters," he says. Not just clams but also tube worms and thick mats of bacteria, white or bright orange. Years later Suess would see those same mats draping the mounds at Hydrate Ridge, with the clams huddled around their edges.

In the meantime, cold-seep communities had been discovered all over the world. They remained a mystery. As Boetius puts it, "Where does it all come from?" In most of the big wide cold ocean, life on the sea floor is much sparser than it is at hot springs and cold seeps because it is sustained only by scraps of organic matter falling from the sunlit surface waters — by trickle-down photosynthesis. At hot springs on midocean ridges, life is sustained by chemosynthesis; the energy source is not the sun but hydrogen sulfide made by hot water flowing over volcanic rock. Sulfide-eating bacteria use that energy to make carbohydrates and build tissue, and the animals either eat the bacteria or incorporate them as symbionts in their tissues. The clams and tube worms at seeps do likewise, but there is no heat source, so the mystery remains: where does all the sulfide come from?

Another related question presented itself to Boetius at Hydrate Ridge: where does all that methane go? Most of the methane made by microbes in the deep sediment layers never gets to the surface. For decades geologists have reported that methane concentrations decrease as sediments become more shallow. Meanwhile, sulfate does just the opposite: it's abundant in seawater, but its concentration in sediments decreases steadily with depth. Invariably, there is a layer where both compounds dwindle rapidly — as if they were both being used up by the same process.

Geologists had suggested one process: maybe there were methane-eating microbes in that sediment layer — microbes that were oxidizing methane with sulfate rather than oxygen because at that depth there is no oxygen. The trouble was, microbiologists had never been able to find such microbes. Friedrich Widdel, a microbiologist at the Max Planck Institute, looked in all sorts of anoxic sediments, from tidal flats to sewage sludge; he would take a sample, add methane to it in the lab, and see if anything ate the methane. "Nothing ever happened," says Boetius. "One postdoc — he's still at Max Planck — he tried for five years. It happened to all of us. I know a hundred microbiologists who tried." Not surprisingly, some concluded that the mystery microbes didn't exist.

But Boetius had a strong reason to believe. It was the other mystery, the one that had brought her to Hydrate Ridge in the first place — all those sulfide-eating clams and bacterial mats squatting on a methane vent. In theory, the missing microbes might solve that mystery too: in using sulfate to oxidize methane, they should be reducing the sulfate to hydrogen sulfide. Nor did this complicated chemical transaction necessarily have to be going on in a single microbial cell. In 1994 Tori Hoehler of the NASA Ames Research Center had suggested that two separate species of microbe, a methane eater and a sulfate reducer, might be collaborating symbiotically in the deep-sea mud.

When Boetius got her Hydrate Ridge mud back to the lab in Bremen, she quickly saw that this idea was promising. She measured the rate at which sulfate was being reduced in the mud and found the highest rate ever measured in marine sediments. She stained the sediments with a dye that would cause microbial DNA to fluoresce under her microscope so she could count the number of microbes in the mud. She found up to 3 trillion cells in an ounce. Normal marine sediments have a mere 30 billion.

Then she tried to identify those microbes using a technique called FISH, for fluorescence in situ hybridization. She had learned it in the lab of the molecular geneticist Rudi Amann. In FISH, the fluorescent dye is attached to a genetic probe that binds only to the equivalent gene in a certain kind of microbe. By using a series of probes, a researcher can get increasingly specific about which microbes are in sediment — as long as someone has made probes for them. Boetius was fortunate to have a probe made by Ed DeLong of the Monterey Bay Aquarium Research Institute, who along with the German researcher Kai-Uwe Hinrichs had recently identified the genes of a possible methane eater in sediments off California.

Boetius's big moment happened in a dark room at the Max Planck Institute late on a Saturday night, when she would rather have been out dancing. She was sitting in front of the microscope, staring at images of sediment slices. She had used probes designed to detect twenty different species of sulfate-reducing bacteria before she finally found one that would produce glowing dots on the screen. The probe from DeLong and Hinrichs, on the other hand, had worked right away: the Hydrate Ridge sediments were loaded with their methane eater, which is not a bacterium at all but a species of archaea, an ancient group of microbes that diverged from bacteria billions of years ago and are as distinct from them now, genetically speaking, as humans are.

Boetius was trying to count those dots on her computer screen, first the archaea and then the sulfate reducers, to find out how many were in her sediments. The cells were clumped together, which made them maddeningly hard to count. Then, in her growing irritation, she noticed something. "I see these stupid clusters of archaea, and now I see these stupid clusters of sulfate reducers," she says. "And they had a very funny shape. The archaea looked like real clumps — lots and lots of cells sitting together. The sulfate reducers were like shells, a circle of sulfate reducers with nothing in the middle. And, really, I sat there for two hours before it finally popped into my head."

The sulfate reducers were stuck to the archaea, forming a shell around them. Tori Hoehler's idea of a microbial consortium suddenly looked compelling. Boetius and Widdel and graduate student Katja Nauhaus at Max Planck later performed the same experiment they and other microbiologists had done so many times before in vain — injecting methane into the sediment. But this time it vanished, and sulfide appeared in its place. Each clump was less than one-thousandth of an inch across and contained hundreds of cells. There were about 900 million clumps in every ounce of sediment at Hydrate Ridge.

The archaea in Boetius's clumps were close relatives of other archaea that live a quarter or a half mile down — the ones that make the methane in the first place. The methane makers assemble the gas from hydrogen and carbon dioxide; the methane eaters do something like the reverse — but not quite, because they don't seem to give off hydrogen. In some way that remains unclear, they pass energy onto the sulfate reducers that surround them. What the archaea get in return is also not clear. "There's some kind of delicate interaction that we do not understand," says Widdel. He has a postdoctoral student and a graduate student trying to grow the consortium in a laboratory, knowing that the reason he and other microbiologists failed to do so in the past is that the microbes grow extremely slowly. "We know it will take time," Widdel says. "We might need two or three Ph.D. theses."

Why is it worth the trouble? Boetius and her colleagues have found the consortium in twenty or so other places around the world — everywhere they have looked, including a mud volcano in the Arctic Ocean, and at cold seeps and hydrate mounds in the Gulf of Mexico. Two years ago, in a kind of crater off the Democratic Republic of the Congo, 10,000 feet down, a team led by Myriam Sibuet of the French Research Institute for Ocean Exploitation discovered a spectacular cold seep with a vast field of clams and mussels, blue shrimp, purple sea cucumbers, and six-foot-long tube worms growing in bushes next to mounds of gas hydrate. Boetius's microbes were in the mud there too. Boetius thinks her consortium, or something like it, provides sulfide at cold seeps everywhere. It is at the base of the food chain for these sea-floor oases.

In the Black Sea, on the other hand, the consortium *is* the food chain. The mats on the sea floor there, and the walls of the chimneys, are a thick patchwork of methane-eating archaea and sulfate-reducing bacteria. The carbonate cores that allow the chimneys to stand tall are a by-product of the microbes' metabolism. (At Hydrate Ridge there are giant slabs of carbonate.) A network of microscopic channels allows water to circulate through the chimneys, supplying the microbes with the chemicals they need. "It's astonishing that as microorganisms they build up structures like that," says Seifert.

It's astonishing, too, to think what Earth would be like if these microorganisms didn't exist. All the methane that is now being converted to carbonate and biomass would instead be bubbling freely from the sea floor — everywhere. Hinrichs and Boetius have estimated that an additional 300 million tons a year of methane would escape from the mud. It's not clear how much of a greenhouse effect that would produce, but it's a good bet that Earth would be a lot warmer — much as it would be, say, if there were no plants drawing carbon dioxide out of the atmosphere. "Everybody knows that our planet would be a different planet if there weren't any plants," says Boetius. "But nobody has thought about who keeps us from having a methane atmosphere." All of which is a lot more than a thought experiment — because in the past the microbes have not always succeeded as well as they do today.

Just off the coast of Norway there is a thousand-foot-high rock cliff that fishermen long ago named Storegga. The name means "big edge." Storegga is at the edge of the continental shelf, where the bottom drops out on fishing boats' depth sounders. On the sea floor here, the bottom dropped out about 8,200 years ago in a massive landslide.

The landslide was one of the greatest in Earth's history. It probably started on the middle of the slope, in a layer of weak, porous sediment, says Jürgen Mienert, a marine geophysicist at the University of Tromsø in Norway. It then propagated rapidly back toward the shelf, as if the slope were a hamburger and a giant from the abyss had taken bites out of it. They were big bites: blocks of mud perhaps 20 miles long, a couple of miles wide, and 150 feet high rushed down the slope. More than 1,000 cubic miles of sediment and rock shifted. The slide ran out 500 miles to the northwest, north of Iceland, where it met the Mid-Atlantic Ridge and was diverted south. Over an area of 35,000 square miles, the whole ocean was as muddy as the Mississippi after a storm, and the sea floor was wiped clean of whatever lived there. "The whole slide happened in a very short time," Mienert says. "Perhaps in a few weeks, perhaps a few days, perhaps a couple of hours."

At the edge of the shelf, where the giant hit rock too hard to bite, there is now a cliff — Storegga. The landslide made a hole in the ocean into which water rushed down, then bounced back. The disturbance propagated a tsunami that flooded coastal areas as far away as Scotland. Along the coasts of Scotland and Norway, the wave ranged between 20 and 50 feet high. It may have crested at 65 feet as it roared up narrow Norwegian fjords.

To say that the cataclysm was caused by deep-living microbes would be excessive — but their methane may have had a lot to do with it. There are gas hydrates under the sea floor at Storegga today, and before the slide, says Mienert, there were probably a lot more. Around 11,000 years ago, as the last ice sheets retreated from Norway and the Norwegian Sea, Atlantic water flowed in and warmed the bottom by about 9 degrees Fahrenheit. Mienert's team has calculated that it would have taken about 3,000 years for that warming to propagate down through the sediment to the base of the hydrate stability zone, where methane is always on the edge of becoming gas. Right at the time of the Storegga slide, the hydrate was being melted by global warming.

On the continental shelf just north of the headwall of Storegga, there are cracks in the sea floor, maybe 3 miles wide and 50 feet deep and dotted with round pockmarks. That's where the methane came out. Frozen hydrate cements sediments together. Melted hydrate, with gas bursting to get out, weakens them. There was already a layer of weak sediments on the Norwegian continental slope, and it is probable that an earthquake was what triggered the slide. But it's equally probable that the gradual melting of the hydrates made it possible — and made it worse.

And it's likely to happen again, somewhere. In the spring of 2000 a team of American researchers caused a tremendous stir when they announced that they had found cracks and pockmarks, much like the ones off Norway, at the edge of the continental shelf near Cape Hatteras. So far they have no way of computing the risk of an undersea landslide there, and thus the possibility of a large tsunami submerging the mid-Atlantic seaboard — although it's surely less imminent than the next major hurricane. They are, however, confident that those cracks and pockmarks are places where methane has exploded from the sea floor.

Pockmarks are turning out to be extremely common, especially in the Arctic; Mienert has seen fields of them in the Barents Sea. Finds like that, along with sediment cores and ice cores that show how the amount of methane in the atmosphere and ocean has fluctuated dramatically in the past, have led to a slew of "methane burp" theories. Huge quantities of methane, the theories say, have escaped from sea-floor hydrates at various times in the past, wreaking havoc.

One of the worst catastrophes in Earth's history, for instance, happened 250 million years ago at the end of the Permian Period, when something wiped out most of the animals. Many researchers blame an asteroid impact, but geologist Gregory Retallack of the University of Oregon has suggested that methane burps from below the sea floor produced a "postapocalyptic greenhouse" that drained oxygen from the atmosphere, leaving animals gasping. If that had happened, it would have reduced the oxygen concentration at sea level to what it is at 16,000 feet today. Gregory Ryskin, a chemical engineer at Northwestern University, favors an even more dramatic scenario. So much methane accumulated in stagnant seas at the end of the Permian, he argues, that when it finally erupted, it ignited, setting most of the planet on fire. Even marine organisms weren't safe; the rising methane brought up anoxic water from the deep, suffocating them. Ryskin thinks we should be keeping a close watch on places like the Black Sea.

Although not many researchers are as concerned as he is, there is evidence that methane escaping from hydrates might have affected climate a lot more recently than the Permian. If so many pockmarks are visible on the sea floor today, it is because they are relatively fresh and haven't yet been filled by sediment washing down to the sea from rivers. The pockmarks date from the last ice age. Submarine landslides, like the one at Storegga, seem to have occurred frequently as the last ice sheets receded, as well as at the end of previous glaciations. Marine geologist James Kennett of the University of California at Santa Barbara has proposed that bursts of methane from sea-floor hydrates were synchronous with, and largely responsible for, virtually all the warmings the planet has experienced over the last 800,000 years. Changes in ocean currents, Kennett says, triggered the methane bursts by channeling warmer water over continental slopes, as at Storegga.

Of the many questions that cling to scenarios of methane-driven climate change, the biggest is this: can methane from melting hydrates actually make it from the sea floor to the atmosphere?

"I would argue that there's zero evidence for that," says Gerald Dickens, a leading expert on sea-floor hydrates and their role in climate change. Zero is a bit strong — Suess and Bohrman have seen blocks of hydrate floating on the sea surface off Oregon. And in the Gulf of Mexico, a team led by Ian MacDonald of Texas A&M at Corpus Christi has observed gas bubbles coated with oil rising to the surface, leaving slicks that are visible on satellite images. But so far, as Dickens says, no one has shown that quantities of methane large enough to change the climate would reach the atmosphere. For one thing, methane bubbles, when they are not coated with oil, tend to dissolve in seawater. For another, aerobic bacteria in the water consume methane. In Dickens's view, what would make it into the air is the carbon dioxide that the methane eaters give off.

CO_2, of course, warms the planet, just not as sharply as methane. Dickens himself proposed the most widely believed case for a climate change involving hydrates. He says it probably happened at the end of the Paleocene Epoch 55 million years ago. Not much that we care about died then; the most prominent victims were deep-ocean microscopic foraminifera that live on the sea floor. But in their tiny shells of carbonate those forams preserve evidence of what happened. Their carbon-isotope ratio shows that all of a sudden there was a lot more light carbon-12 — the kind that living organisms favor, the kind in sea-floor hydrates — in the water around the forams.

If roughly 2,000 gigatons of methane came out of hydrates then, Dickens calculates, it would explain the isotope excursion. The excursion has been detected in sediments all over the world, and

that's how Dickens pictures the increase in methane emissions —
fifty cold seeps wherever there now is one, all over the world. Even
before the hydrates melted, he says, the planet had been warming
for a long time, by about 7 degrees Fahrenheit, which melted the
hydrates. The methane releases added another 2 or 3 degrees. In
other words, their impact was not catastrophic, but it was not trivial
either.

Today people are warming the planet by putting carbon dioxide
into the atmosphere — and methane too. There are methane-gen-
erating archaea in our rice paddies, for instance, and also in the in-
testines of cows. Might we be about to trigger some serious burps?
Euan Nisbet, a geologist at the University of London, points out
that the Arctic, where the warming is expected to be strongest, is
vulnerable — both on land and in shallow seas there are hydrates
that are stabilized mostly by low temperatures rather than by high
pressures. The methane there "would probably take some decades
or centuries to come out," he says. "But once it started, it would be
essentially unstoppable."

Nisbet and virtually every other researcher say there is no need
to panic — but everyone agrees it would be reassuring if we understood a little better how carbon cycles through our planet. The
problem, Dickens says, is that we have been ignoring a large part of
the cycle. We've been assuming it stops at the sea floor, that organic
matter buried there by the steady snow of sediments is removed
from the cycle forever. But it isn't. Archaea deep in the mud con-
vert it into methane, and the methane ends up in giant frozen reservoirs, and some of it leaks back into our ocean and maybe into
our atmosphere. The reservoirs are always changing, always waxing
and waning, always charging and discharging; Dickens compares
them to electric capacitors. He compares them also to more familiar carbon reservoirs.

"It's almost like you've got to think of them like forests," he says,
"where you have photosynthesis and respiration and trees growing
and expanding and dying — there's always this carbon turnover
through the biomass. Well, it's the exact same sort of thing. It's this
big carbon storehouse, but it's all in methane, and it's all controlled by the microbes." And there's the rub: if methane hydrates
are like forests, then we don't understand the trees.

*

In 1987, in a laboratory of the Scottish Marine Biological Associa-tion in Oban, Scotland, geologist John Parkes opened a series of twenty-ounce cans that had been shipped to him from Peru. Inside each was a two-inch-thick disk of mud. Parkes was dressed warmly, because he was working in a chilly room at 60 degrees Fahrenheit. Geochemists on the drill ship, on the expedition led by Suess, had cut slices from several drill cores with sterile knives, capped them, flushed them with nitrogen to banish all oxygen, sealed them in cans, and sent them to Oban.

Parkes opened the cans and extracted tiny cores from each. In those tiny cores he found that he could detect bacterial activity — the reduction of sulfate to smelly sulfide, for instance. Staining minute samples of sediment with a fluorescent dye that binds to DNA, he found he could count cells under his microscope, just as Boetius would do with her Hydrate Ridge samples many years later. At the time no one had done this with drill cores. And when all of it was done, says Parkes, "we'd discovered 10 percent extra biomass on Earth." More recent estimates place it at more like 30 percent.

Parkes's peers did not rush to anoint him. For decades geochem-ists had been reporting that sulfate and methane were mysteriously vanishing in sediments. For decades, too, petroleum geologists had been reporting that some kind of microbe seemed to be chewing up oil in deep reservoirs. "These facts were staring us in the face, but nobody put it all together," says Parkes. When he did, it was suddenly controversial. People suspected that his samples were contaminated by surface bacteria. They refused to see the whale.

They see it now. Two years ago the drill ship *Joides Resolution* went back to the same waters off Peru where Parkes's first samples had been taken. This time Parkes was on board, along with a dozen other microbiologists and a few geochemists, including Dickens. It was the first expedition specifically designed to study the deep bio-sphere. And the first result was simply to confirm what Parkes had been saying all along. "There are masses of bacteria down there," says Bo Barker Jørgensen of the Max Planck Institute, who was co-chief scientist of the expedition.

The researchers found microbes in all the sediments they exam-ined. There were more under the coastal waters of Peru than in the open Pacific; more near the sea floor than 1,400 feet below it. But there were intact microbes everywhere. In the upper layers, typically, they were reducing sulfate; in the lower ones they were making methane; and in between they were oxidizing methane.

The existence of the deep biosphere is established — but it remains an astonishing paradox. "From all we understand about the energy requirements just to stay alive, it's much higher than the energy they have," says Barker Jørgensen. If the deep microbes spend as much on maintenance as surface microbes do, he says — repairing radiation damage to their DNA, keeping their membranes intact — they should have nothing left for the microbial prime directive: divide and multiply. Barker Jørgensen's expedition looked for some new energy source in the sediment, some exotic new combination of fuel and oxidant, and found none.

Parkes thinks the microbes' secret is their slowness: "These things are dividing every thousand, ten thousand, hundred thousand years. There's nothing to eat them; bacteria near the surface have to grow fast because they get eaten by protozoa and ciliates, but we've not detected those kinds of organisms in the subsurface. So bacteria there can concentrate on maintenance, rather than wasting energy on division." And yet they must have lived long enough and divided often enough and mutated often enough to evolve through natural selection, because they are well adapted to their environment. Parkes has found microbes in deep sediments that grow best at precisely the pressure at which he found them. "They are responding on geological time scales," he says. "That's the fascinating thing."

Microbes living under the sea floor today, Parkes speculates, may have survived the growth and splintering of continents, the opening and closing of oceans; they may have been buried, subducted, frozen in hydrate, and spat out of a mud volcano, only to be buried, subducted, and spat out again. While we were waiting for our evolutionary fast lane to be paved, racing through all of human prehistory and history in the time it takes one of them to divide once, they have been living in time with the planet's deepest, slowest rhythms. They have been living almost like rock, which is precisely what made them so easy to miss. They have always been there, from the deepest past, but only now have they finally penetrated into our awareness. Given their collective influence, it's about time.

WILLIAM LANGEWIESCHE

A Two-Planet Species?

FROM *The Atlantic Monthly*

IN THE AFTERMATH of the breakup of the space shuttle *Colum-bia,* an important debate on the purpose and future of the U.S. hu-man space flight program is under way, though perhaps not as forthrightly as it should be. The issue at stake is not space explora-tion in itself but the necessity of launching manned (versus ro-botic) vehicles. Because articles of faith are involved, the argu-ments tend to be manipulative and hyperbolic. If the debate is to be productive, that needs to change.

The proponents of manned space flight — particularly at NASA — are in an unenviably defensive position. They argue weakly for the operational flexibility provided by astronauts in space and trot out the story of the Hubble Space Telescope's in-orbit repair, as if this single success might justify two decades of shuttle flights; they advertise the applied scientific benefits of performing laboratory-style experiments in orbit, though they cannot point to much of significance that has arisen from them; and finally, bizarrely, they argue that only the excitement of a human space flight program can persuade the American public to foot the bill for the robotic ef-forts — a self-indicting logic if ever there was one. When all else fails, they fall back on the now empty idea that the shuttle program is a matter of national prestige — or on the still emptier claim, made obliquely by NASA administrator Sean O'Keefe last spring, that a retreat from human space flight would be akin to a return to the Stone Age for all humankind.

That sort of talk allows the opponents of human space flight to play the role of the rationalists. They question the effect on the national psyche of successive accidents, debunk NASA's orbital science, and describe the ultra-expensive International Space Station as hardly more than an artificial destination for an ill-conceived spaceship with nowhere else to go. They are right about all that. While they are at it, they juxtapose the human space flight budget (currently about $6 billion annually) with underfunded research on Earth (for example, in oceanography and clean energy), and they point to the history of absurd cost overruns in both the shuttle and the space station programs; NASA's manipulation of budget figures; and its well-demonstrated managerial incompetence. Again, these are all valid arguments. But the critics here are not merely noting the problems in human space flight; they are setting up a straw man — the shuttle — in order to knock down a much larger thing. An honest national debate would demand more.

Such a debate would almost certainly lead to the conclusion that the United States has for thirty years followed human space flight policies that are directionless and deeply flawed and that those policies must now be radically changed, with whatever regret about the historical costs. Chances are that such a change would involve maintaining the space station in its current unfinished state, and doing so in the least expensive way, with cargoes and minimal crews launched atop subsidized Russian rockets — after laws that currently prohibit such subsidies are repealed. But it would also involve permanently grounding the shuttle fleet, and shutting down much of the associated NASA infrastructure. The savings would be large and immediate.

This would not mean, however, that the opponents of human space flight had won. Indeed, it may be that a pause to regroup is precisely what a vigorous human space flight program now needs. One thing for sure is that the American public is more sophisticated than the space community has given it credit for. In the event of a grounding the public might well be presented with a question now asked only of insiders — not whether there are immediate benefits to be gleaned from a human presence in space but, more fundamentally, whether we are to be a two-planet species. If upon due consideration the public's answer is "yes," as it probably should be, the solutions will be centuries in coming. Compared with the scale of such an ambition, a pause of a few decades now to rethink and rebuild will seem like nothing at all.

BILL McKIBBEN

Crossing the Red Line

FROM *The New York Review of Books*

1

IN OCTOBER 2003, two consultants on energy, one of them the former head of planning for Royal Dutch/Shell, produced a report for the Pentagon that they titled "Imagining the Unthinkable." The twenty-two-page document outlined one possible sequence of events that could result from the greenhouse effect. First, melting Arctic ice, less dense than salt water, would flood the Atlantic with cool fresh water. By 2010 it would slow down or shut off the local branch of the relatively warm circulatory system that we know as the Gulf Stream. As a result, while the rest of the globe continued to grow warmer, western Europe and eastern North America would turn sharply colder, and the relatively dry interior of Europe soon would have a climate comparable to present-day Siberia.

These developments would lead to shortfalls in grain harvest felt around the world. Before long, wars would threaten to erupt over "desperate need for natural resources" rather than "over ideology, religion, or national honor." Every time "there is a choice between starving and raiding, humans raid," the authors write. And so, as the number of human beings outweighed the greatly reduced carrying capacity of the planet, "deaths from war as well as starvation and disease" would over time have to "re-balance" our population.

The report is not novel — such visions of rapid and violent climate change have become common in recent years as our understanding of the wild weather swings in the climatic temperature record has grown. Nor is it extremely unlikely — a series of recent papers in *Nature* have noted just the sort of freshening of Atlantic waters that could set off this particular sequence. To focus on the report's more lurid predictions, however, would be to miss the real point, which is the mere fact of its existence. In an administration that has refused to even acknowledge global warming as a serious human problem — an administration whose Environmental Protection Agency removed the section on climate change from its annual report to avoid offending the White House — the report's calm and straightforward acceptance of the basic laws of physics and chemistry seems remarkable. "There is substantial evidence to indicate that significant global warming will occur during the twenty-first century," the summary begins. "Alternative fuels, greenhouse gas emission controls, and conservation efforts are worthwhile endeavors," it concludes, before adding a list of other, more exotic, responses that will likely have more direct appeal to the Pentagon ("explore geo-engineering options that control the climate").

Though the Pentagon officially played down the report (a spokesman promised it would not be passed along to Defense Secretary Rumsfeld), its internal think tank, the Office of Net Assessments, made no attempt to hide it from public view. In fact, someone made sure to pass it along to *Fortune* magazine, which published a long account of its contents. It suggests that even some in official Republican Washington have begun to notice how out of step the United States has become on this issue — after all, every other government in the industrialized world (not to mention virtually all of organized science) has declared that climate is changing swiftly and dangerously. In those capitals, "alternative fuels, greenhouse gas emission controls, and conservation efforts" are not just "worthwhile endeavors" but cornerstones of national policy.

If the Pentagon wished, it could envision similarly disastrous consequences of global warming. The Antarctic shelf could collapse. Water shortages could result from increased evaporation. Millions of refugees could flee from low-lying lands. The interiors of continents could turn into deserts, while mosquito-borne diseases could rapidly spread. In view of the scale of the issues, it is worth surveying recent proposals to hasten the transition away from fossil fuels. It is not, on the whole, an encouraging picture.

*

Two of the books under review, *Bush Versus the Environment* and *Strategic Ignorance,* include climate and energy policy as one example of many in a catalog of administration environmental follies. But it must be said that criticizing Bush's policies on the environment is depressingly easy to do. For more than three years now, day after day and week after week, a small circle of political appointees at the EPA, the Forest Service, the Interior Department, and the Department of Agriculture have proceeded methodically to wreck the system of environmental oversight that dates back to the Nixon administration. Apart from their silence on global warming, they have overturned rule after regulation, largely ceased enforcement actions concerning pollution of the atmosphere and water, and reined in inspectors. Their work is not inspired by a grand ideological vision — it's not like Bush's foreign policy, say, with its idea of America dominating the world. Instead it's institutionalized corruption: a steady payback to the logging, mining, corporate farming, fossil fuel, and other industries that contributed heavily to put Bush in power. The scale of this assault on the environment is so large as to be numbing. With a hundred battles occurring simultaneously and without a majority in either chamber of Congress to hold hearings or issue subpoenas, the environmental movement has been almost paralyzed. In Congress and the administration, loss has followed loss in such steady succession that even the most conventional environmentalists, usually bipartisan to a fault and reluctant to jump into electoral politics, now find themselves with a single goal: defeating Bush in November.

Some of the recent books are designed to serve as ammunition for that battle. Robert Devine's *Bush Versus the Environment* is full of reports of the harm the administration is causing to woods, coalfields, and tundra. But it is Carl Pope's volume, written with the editor of *Sierra* magazine, Paul Rauber, that makes the strongest case. Pope has spent most of his life in environmental politics — the politically shrewd executive director of the Sierra Club since 1992, he's clearly fed up with polite dissent. He has written a splendidly fierce book, especially so in its account of the public relations effort to limit the political damage from Bush's antienvironmental policy. The archfiend in this account is Frank Luntz, the veteran Republican pollster whose surveys showed that most Americans wanted no part of the president's plans for the environment. In response, Luntz suggested the words and images that might most effectively hide the truth. The bill that turned the national forests back to loggers in the name of protecting against wildfire, for instance, was called the "Healthy Forests Initiative," though, as Pope suggests, "Horizontal Forests" would be more accurate. (Others have suggested "No Tree Left Behind.") The bill to permit continued high levels of mercury and sulfur pollution was styled "Clear Skies."

Luntz told Bush to stop using the phrase "global warming" (in a leaked memo, he stressed that "while 'global warming' has catastrophic connotations attached to it, 'climate change' sounds a more controllable and less emotional challenge") and to emphasize the (false) statement that there is no consensus among scientists on the issue. "Should the public come to believe that the scientific issues are settled, their views about global warming will change accordingly," Luntz wrote. "The scientific debate is closing [against us] but not yet closed."

Bush has evaded energy and climate issues, but Bill Clinton and Al Gore weren't conspicuously better. That's because dealing with global warming is not a matter of simply paying a relatively small price to clean the air or water. It will demand nothing less than the overhaul of the entire global economy, which is currently based on the very fossil fuels whose combustion we can no longer afford, but whose replacement remains technologically, economically, and politically more challenging than perhaps any transition in modern human history.

There are those who think this shift won't be terribly hard — many people harbor the hopeful belief that only the political power of the big oil interests prevents the swift transition to a world of solar and wind power and, especially, the widespread use of electrochemical devices called hydrogen fuel cells to produce energy. Hydrogen fuel cells seem promising because they can produce electricity and heat from the chemical energy contained in hydrogen, an abundant element. In most near-term plans, the hydrogen would come from hydrocarbon fuels, most likely natural gas, which would mean continued release of greenhouse gases. Over the longer term, we might be able to build enough renewable energy capacity — solar and wind in particular — to "electrolyze" water, stripping away its hydrogen in a truly renewable fashion.

Among the true believers in this future is the business-friendly energy and environment editor of *The Economist*, Vijay Vaitheeswaran, whose *Power to the People* splits its cover between a picture of spinning windmills and another of a field of sunflowers. He argues that market forces (particularly electricity deregulation), new environmental laws, and most of all a wave of technological innovation will soon combine to produce an energy revolution "bigger than the Internet." In newsmagazine tradition, Vaitheeswaran writes with fervor about every development he comes across. Some academics, he writes, "are working on a . . . radical concept called biolysis that involves manipulating the metabolism of algae and other life-forms to release large quantities of hydrogen when they decay." This accounts for the title of a subsection of his book, "Pond Scum to Petro-Hydrogen." His reporting is at its most emotional when he visits "Plant 1, the top-secret bunker from which Ballard [the leading hydrogen fuel-cell manufacturer] intends to conquer the world." In so doing, "I became the first journalist ever to penetrate this Fort Knox of the fuel-cell world," and there he witnessed "dramatic breakthroughs," "visionaries" hard at work, and long lines of "brilliant physicists and electrochemists."

For all his gush, however, some of his conclusions are sound: we need, he writes, a carbon tax that would encourage investment in new low-polluting technologies. Those technologies won't include the excessively expensive option of nuclear power; instead, eventually we will live in a world of "micropower," where small-scale power plants use first natural gas and then renewable sources like wind and sun to turn hydrogen into plentiful and cheap energy. This proposal mirrors the European consensus that has emerged in recent years; the fact that it is being espoused in the United States by a member of Europe's business establishment should make it more palatable to Americans. As he titles his epilogue, "The Future's a Gas."

Indeed, you could say that even the president has signed on to parts of this program. Not the carbon tax, certainly, but in his most visible bid for an environmental accomplishment of some kind, the president committed more than a billion dollars to the future search for a hydrogen-powered vehicle, something he (doubtless with his pollster Luntz's help) dubbed the FreedomCar. Bush said,

"With a new national commitment, our scientists and engineers will overcome obstacles to taking these cars from laboratory to showroom, so that the first car driven by a child born today could be powered by hydrogen, and pollution-free." Vaitheeswaran is, as ever, enthusiastic. But many observers believe the president signed on to hydrogen as a cheap way to avoid any current changes. Forget increasing automobile efficiency right now; instead, promise something shiny far enough in the future (2018 or so, to go by his "child born today" theme) that it won't affect today's taxpayers (or fossil fuel interests). So the question becomes: how realistic is the promise of a near-term hydrogen future?

Luckily, a pair of new books help unravel that question. The most direct answers come from Joseph J. Romm, in *The Hype About Hydrogen*. Romm is no opponent of hydrogen. In his years in the Clinton administration Department of Energy he helped to strongly increase funding for hydrogen research; later, as an investor, he helped back Sure-Power systems, which in 1999, amid much publicity, sold the first commercial stationary hydrogen fuel cell to an Omaha-based credit card processor that was willing to pay a premium for its promise of uninterruptible power for its office buildings. This is the kind of fuel cell–based micropower installation that Vaitheeswaran foresees as a growing source of energy; sadly, since 1999 SurePower hasn't made a comparable sale, and the company that actually manufactured the fuel cells is now phasing them out. And such on-site installations are the easy part of the fuel cell future — putting them in cars will require making them small and finding a way to distribute hydrogen around the country.

Such technical obstacles, described in great detail in Romm's flatly written book, have turned him into a hydrogen realist. He considers, for instance, all the different methods of distributing hydrogen around the country for use by drivers of Bush's FreedomCar. Suppose you load compressed hydrogen into canisters and put them on the back of tractor trailers: you will need fifteen of these trucks to serve the same number of vehicles as one gasoline tanker does today, and if on average they're traveling three hundred miles, they're using 40 percent of all the energy they deliver just to transport it. Suppose you decide instead to produce the hydrogen at filling stations with small steam methane reformers. A recent study by Argonne National Laboratory, making fairly optimistic technical predictions, found that building the necessary infrastructure to service 40 percent of American vehicles by this method would exceed $600 billion.

But here's the really startling piece of news that emerges from Romm's book: if you worry about global warming, that investment won't necessarily get you much. For the last couple of years I've driven a hybrid-electric Honda Civic that has a gasoline engine and also a self-recharging electric motor to augment power. It works exactly like a normal vehicle, goes eighty miles per hour on the highway, drives easily up any hill, and gets fifty miles to the gallon without any of the problems of cost, technology, or infrastructure that hydrogen presents. Even if we manage to solve the huge number of difficulties that Romm invokes, in the end the hydrogen car that results will almost certainly use natural gas as the feedstock for obtaining hydrogen. It will produce about as much carbon per mile as my Honda Civic (which, in turn, should only get better with fifteen years of tinkering). Even diesel engines, which are as old-fashioned as automotive technology gets, may provide comparable reductions for far less money. Romm concludes: "Hydrogen vehicles are unlikely to achieve even a five percent market penetration by 2030 . . . Neither government policy nor business investment should be based on the belief that hydrogen cars will have meaningful commercial success in the near- or medium term."

For now, at least, the smart money seems to agree. Ballard's share price has fallen from a high of $140 to as low as $6, and 25 percent of its workforce has been laid off. Paul Roberts, in his book *The End of Oil,* quotes a hydrogen engineer: "I'm afraid that when we finally get people to stop associating hydrogen with bombs and the Hindenburg explosion, the next word they'll think of will be 'scam.'"

The End of Oil is a stunning piece of work — perhaps the best single book ever produced about our energy economy and its environmental implications. Paul Roberts writes regularly for *Harper's Magazine,* and he has schooled himself deeply in these questions. He writes with authority and depth, which makes it all the worse that his message is almost unrelievedly gloomy — the dismal prospects for hydrogen are just a small part of his argument. He begins with a careful examination of an important question: are we about to run out of oil? This, of course, is a question people have been asking since the oil shocks of the 1970s, but now they are asking it with greater and greater urgency because we've used vast quantities of oil in the three decades since, more oil than prospectors have been able to discover. (Earlier this winter, in fact, Shell's CEO resigned when it became apparent that the company's reserves had been overstated by billions of barrels.) Country after country — Britain, Mexico, Nigeria, Norway — seems to have neared or hit the peak of its production, and even with all kinds of new technological tricks their oil wells can no longer maintain output. Even the mighty Arab fields are not eternal. Roberts begins his book with an account of his shock on hearing that the Saudis' vast wells at Ghawar now need huge injections of water to keep pumping.

Meanwhile, demand continues to soar. Roberts's statistics are up to date, and appalling. As recently as last summer, forecasters were predicting world oil demand would rise 1.3 percent this year. But the Chinese desire for automobiles is growing so rapidly that in fact demand may rise 2.25 percent this year — or two million barrels more a day. He quotes the Tufts researcher Kelly Sims-Gallagher, who attended the recent Seventh Beijing International Auto Show. "You had no elbow room — you were just being carried along on this wave of people," she said. "I've been in China many times, but I had never experienced crowding on this scale, and that was when it hit home how many people there are who are absolutely serious about buying a car." Or, in the words of a Chinese auto executive, "You could see consumers' fever."

Roberts also concentrates on natural gas, which many see as the "bridge fuel" that will help us move from fossil fuels to an age of renewable energy. So many see it that way, in fact, that heightened demand for natural gas, largely from new electric power plants, is already straining supply. Unlike oil, gas is hard to move around: getting it from distant wells to North America or China will require liquefying it, an expensive technology whose spread will take time. And though it's cleaner than coal or oil, it still produces carbon — at best it's a stopgap.

2

In the long run what we need are carbon-free fuels. There are several possibilities. Some in the coal industry hope they can figure out ways to capture the carbon dioxide that comes from burning coal and inject it safely into underground pockets where it can't warm the climate. This technology, as Roberts points out, is still in its infancy, and even the best cases that can be made for it are not exactly simple: "For every freight car of coal delivered to a [power plant], three cars of captured CO_2 would need to be removed and somehow transported to a safe repository and placed underground — a task that, on a global scale, would involve handling a volume of waste material larger than the combined tonnage of the steel and iron industries." Such efforts would add 30 to 50 percent to the cost of the electricity.

Then there's solar energy and wind, the truly renewable sources of energy. In a perfect world, we'd use them to provide the electricity we need for houses and businesses, and also to extract the energy from hydrogen in fuel cells to power our cars. And indeed this corner of the industry is finally starting to really grow. During 2002 alone about $7 billion was invested in wind energy — the market is doubling in size every two and a half years, and by 2020 could easily be supplying 12 percent of global power needs. Solar panels are getting more efficient and less costly by the year as well — enough so that even without government subsidies they will be competitive in sunny parts of the world, including the U.S. Southwest, by 2008.

The problem, says Roberts, derives from the very nature of wind and sun. Sometimes the wind dies; sometimes the sky clouds over. This "intermittency" means that if a utility wants to add 100 megawatts of wind capacity, it has to build 250 megawatts' worth of new wind turbines, which is enormously expensive to do. The ratio for solar power is even worse. Their intermittent effectiveness also means that they lack what the utilities call "dispatchability." Unlike a gas or coal plant, in which you can increase power whenever demand soars, you simply can't count on the renewables. Experts quoted by Roberts say this limits them to providing perhaps 20 percent of a region's power — past that point there are too many power disruptions.

"In the simplest terms, the energy challenge of the twenty-first century will be to satisfy a dramatically larger demand for energy while producing dramatically less carbon," Roberts concludes. "Yet the availability of carbon-free energy on a mass scale will not happen without significant technological developments." Such breakthroughs will not happen "until the market regards carbon as a cost to be avoided — not just in 'progressive' economies like Germany or England, but in the big economies of Russia, China, and above all the United States." But, as we have seen, our leaders are largely uninterested in the carbon taxes that might help change the system.

Roberts's prescription, and it is a balanced and rational one, involves muddling along for the moment while waiting for technological advance. As fears about global warming grow (as more Pentagon reports emerge, for instance), political support for carbon taxes and for taxes on gas-guzzling automobiles will increase. We should speed up the transition to natural gas, which, owing to its lower carbon content, will buy us a decade of grace. And we should hedge our bets technologically, "aggressively pursuing as many technologies as we can afford to" but not locking ourselves into any particular choice until we see where the best results lie.

This more or less describes the efforts of many outside the Bush administration who are trying with some success to influence energy policy, working from state capitals, city halls, and even some corporate headquarters. A number of states have adopted standards requiring utilities to provide 10 or 20 percent of electricity from renewables within the next decade. For instance, Arnold Schwarzenegger has said he will continue to demand higher mileage standards for cars in California. Even in the U.S. Senate, forty-three senators managed to vote for an (exceedingly modest) climate bill sponsored by Arizona Senator John McCain.

This somewhat plodding approach is balanced and rational. It may be our only bet, in view of the power of the fossil fuel lobby and consumer resistance to higher prices. But it's also an extremely dangerous gamble, even more than Roberts realizes. Before the Industrial Revolution, carbon concentrations in the atmosphere were about 275 parts per million. Since global warming was first seen as a danger in the 1980s, scientists have used twice that figure — 550 parts per million — as a red line that should not be crossed.

Beyond it, they warn, the effects on the climate may be impossible to control. Right now, carbon dioxide makes up more than 380 parts per million in the atmosphere, and the proportion is steadily rising. So steadily that some, including Roberts, say it may no longer be possible to stay below 550 parts per million. We should, he writes, be cautious about adopting technology so as not to make costly mistakes, even if it means "temporarily" crossing that border. "Granted, such a strategy is risky," he writes. Still, paraphrasing the Stanford economist Richard Richels, he contends that "by lowering replacement costs and allowing alternative technologies to mature, we dramatically improve our chances for making even greater reductions later on."

In truth, though, the more we find out about climate, the more terrifying such gambles seem. Michael Oppenheimer, the Princeton researcher and longtime prophet of global warming, recently warned about the impact of such a strategy. "Say we go temporarily to 600 parts per million and stay there for a hundred years" (which is roughly the time that carbon dioxide stays in the atmosphere). "We find," he told me,

> it makes a difference of generally around half a degree Celsius. That may not sound like a lot, but the target of a lot of us here is to keep temperature rise to one or two Celsius more than we've already done. So that's a significant chunk. What you put at risk, perhaps, is the stability of the West Antarctic Ice Sheet, the Greenland Ice Cap. If you think the Gulf Stream predictions are scary, wait till the Pentagon starts calculating the sea level rise that comes when Antarctic ice slides into the ocean.

Two new books, in fact, make a strong case that the damage is not some future prospect but instead already very real, if only one knows where to look. *Feeling the Heat*, a collection of pieces by environmental reporters for *E* magazine, looks at regions around the world (including New York City and its harbor, which are also the subject of a big-budget greenhouse disaster movie, *The Day After Tomorrow*, which Hollywood is now releasing). Their reporting from the northern Pacific coast, with its shrinking glaciers, or from Australia, where coral reefs are being bleached, or the ocean off California, where maritime food chains appear in danger of collapse, suggests the claim of the editor, Jim Motavalli, in his introduction to the book, that "whether it is politically convenient or not, 'global warming' has arrived."

Later this month, Basic Books will publish Ross Gelbspan's *Boiling Point*, the long-awaited sequel to *The Heat Is On*, his classic account of how oil and coal companies have been poisoning the debate over global warming with their expensive advocacy of junk science. Gelbspan, more than anyone else, helped to expose the disinformation, only to see those energy companies gain huge new power in the Bush administration. Now he worries that even people who are concerned about global warming, in their dismay at the lack of progress, may be setting the bar much too low, advocating only the smallest and most politically practical changes, all the while ignoring the physics and chemistry indicating that a large-scale shift is likely to take place soon. His concluding chapter, "Rx for a Planetary Fever," sets out perhaps the most thought-out and plausible proposal for rapidly accelerating the transition to a new world.

In my reading, though Roberts is clearly more "realistic" politically, Gelbspan's desperate demand for convulsive change displays a more acute realism, one informed by deep immersion in climate science. In view of the current political impasse over environment issues, it seems to me worth making the special efforts his analysis demands. Not just for a "Manhattan Project" to dramatically speed up technological innovation, but also for large-scale attempts to make real changes in personal behavior. And not just to use energy more efficiently but to use somewhat less of it. Two years ago, when Californians began to fear power shortages, they managed, simply through conservation, to cut electric demand more than 10 percent in the course of a few weeks. These small changes in habit could reinforce emerging technologies. For instance, my wife and I have solar panels on the roof of our house to supply some of our power, and they work well — but their biggest effect is to make us far more conscious of turning off lights. Similarly, my hybrid car saves energy in part because of its brilliantly designed engine but also because it comes with a display that tells me constantly how much gas I'm using and this, as a consequence, has cured me of a heavy foot on the pedal. If Roberts is correct in saying that the development of solar and wind power will face serious problems of "intermittency" and "dispatchability," then we may even need to start thinking about energy a little differently — using more when it's easily available and less when it isn't.

It would, of course, take leadership to make such changes, and one of the greatest sins of the Bush administration is that it squandered the best opportunity for that leadership we've ever had. It's hard to imagine that Americans in the immediate aftermath of September 11 wouldn't have been open to a message of energy conservation and energy transition. In one speech the president could have made the SUV an indulgence to be avoided and the solar panel an almost mandatory accessory for every good patriot. Instead, we were told to return to business as usual. And we have, burning more energy than ever. Now we're three years closer to the eventual reckonings.

James Gustave Speth's *Red Sky at Morning* is a particularly useful summary of the ecological situation. A negotiator for ecological treaties in the Carter administration, he went on to found the World Resources Institute and serve as CEO of the United Nations Development Programme. Now dean of Yale's School of Forestry and Environmental Studies, he has probably spent more hours in international conferences and treaty parleys than any other American. His careful and judicious book concludes that much of that effort has been wasted: "Thus far, the climate convention is not protecting climate, the biodiversity convention is not protecting biodiversity, the desertification convention is not preventing desertification, and even the older and stronger Convention on the Law of the Sea is not protecting fisheries." With the exception of the relatively simple pact to protect the ozone layer, competing national and economic interests have kept most treaties weak: "Global environmental problems have gone from bad to worse, governments are not yet prepared to deal with them, and, at present, many governments, including some of the most important, lack the leadership to get prepared."

Speth offers a long and persuasive list of specific changes in national and international policies that need to be made. But, tellingly, he too ends with "the most fundamental transition of all," a "transition in culture and consciousness." He quotes from a group of researchers at the Stockholm Environmental Institute. Instead of the Pentagon planners' visions of endless raiding wars over food, they imagine a planet where

> preferred lifestyles combine material sufficiency with qualitative fulfillment. Conspicuous consumption and glitter are viewed as vulgar throwbacks to an earlier era. The pursuit of the well-lived life turns to

the quality of existence — creativity, ideas, culture, human relationships and a harmonious relationship with nature . . . The economy is understood as the means to these ends, rather than an end in itself.

It is the true measure of our desperate position that the frail hopes expressed by Speth may turn out to suggest the most solid and practical advice anyone can give.

JAMES McMANUS

Please Stand By While the Age of Miracles Is Briefly Suspended

FROM *Esquire*

A DAMP, NASTY THURSDAY, January 15, 2004. The President's Council on Bioethics is meeting in the downstairs conference room of the Wyndham Hotel, four or five blocks from the White House. Minimal risk-to-benefit ratios for cutting-edge medical research are what the panel is trying to calibrate. Thumbs-up or thumbs-down, live or die. It's 8:45 in the morning, day one of week three of the most crucial election year since 1932, or perhaps 1860. Not to be melodramatic.

To get a better handle on the ethical objections to embryonic stem cell research, I've been listening with as much detachment as possible, given my twenty-nine-year-old daughter's ongoing slow death from juvenile (type 1) diabetes, one of several diseases likely to be cured by this research. Bridget has already undergone a vitrectomy — open-eye surgery to remove vision-blocking blood clots and scar tissue from her vitreous humor — and several rounds of laser treatment to help keep her retinopathy in check. During these procedures, she needs to remain awake while a gonioscope is held against her eye and an ophthalmologic surgeon burns her pigment epithelium about eighteen hundred times with two hundred milliwatts of light. In each eye. Victims of juvenile diabetes can go blind because their elevated blood-sugar levels cause capillaries throughout their bodies, especially in their eyes and extremities, to leak and proliferate in unhealthy ways.

But like I said, I'm trying to listen to every voice here, even that of Alfonso Gómez-Lobo, the dapper metaphysician from George-town who proclaimed half an hour ago in his lilting Chilean ac-cent that "all of us were once a blastocyst." His point was that *no* blastocyst, cloned or otherwise, should ever be destroyed for its cells, however great the possible benefits. I wanted to say that we all were once an ovum as well, yet we don't hold a funeral every time a woman who's made love has her period. That being evolved from amphibians doesn't keep us from deep-frying frog legs and wash-ing them down with Corona. That defenders of animal rights fer-vently believe that eating meat, even fish, is a sacrilege, but we don't let them dictate to Smith & Wollensky or cut off government subsidies to the beef industry . . .

To be fair, the Bush council straddles both sides of the fulcrum: ten members on Gómez-Lobo's side, seven on mine. Each is a bril-liant and well-informed ethicist, doctor, or legal scholar; nearly all have an M.D. or Ph.D or sometimes, as in the case of chair-man Leon Kass, both. Today they're presenting in public the ideas they've already fine-tuned and published.

Kass, Gómez-Lobo, columnist Charles Krauthammer, and the rest of the majority argue that embryonic stem cell research would put us on a slippery slope toward organ farms and cloned children. Their hostility focuses on a procedure called somatic-cell nuclear transfer, in which the nucleus of an ovum is removed or deactivated and replaced with the nucleus from a human donor cell. After chemicals coax the doctored egg to reproduce itself, what forms af-ter five days is a blastocyst, a cluster of one hundred to two hundred cells, including the stem cells prized by researchers. This early-stage embryo could be used for either reproductive cloning (to make a baby that's genetically identical to the donor), which al-most no one favors, or therapeutic cloning (to isolate and harvest its stem cells) to advance the field of regenerative medicine.

Chapter 6 of *Human Cloning and Human Dignity,* published by the council in July 2002, is where its members most fully address the ethics of cloning. The majority recommends, "albeit with re-gret," that therapeutic cloning ought not be pursued. "The cell syn-thesized by somatic-cell nuclear transfer, no less than the fertilized egg, is a human organism in its germinal stage." Somewhat defen-sively, the majority adds, "It is possible that some might suffer in the future because research proceeded more slowly. We cannot sup-pose that the moral life comes without cost."

No one wants cloned babies or fetuses cultured in hatcheries. But the council's minority is willing to let a small number of early-stage embryos — cloned embryos as well as those fertilized in vitro and now stored in the freezers of infertility clinics, bound for the Dumpster if not donated to science — be destroyed during medical research. Why? Because only embryonic — as opposed to adult — stem cells are "pluripotent," which means they're capable of morphing into any kind of cell in the human body. These "differentiated" cells could then be programmed to replace diseased neurons, heart-muscle tissue, or insulin-producing islet cells for diabetics. Genetically compatible with the patient, they would not be rejected by her immune system. Bona fide miracle cures are what we are talking about here. Even before they're perfected, they'll dispense potent medicine — hope. But only if the president's restrictions on this research are lifted.

In the summer of 1979, Bridget was four and a half, a pink-cheeked, blond kindergartner with a respectable forehand and a new baby brother. In the upper right side of her abdomen, though, her immune system suddenly attacked the islets of Langerhans in her pancreas, inexplicably mistaking them for foreign invaders. Within a couple of weeks, almost all of her islet cells had been obliterated. Her body no longer made enough insulin, the hormone that regulates the passage of nutrients into the cells, so she couldn't process food into energy. Long before we knew what was happening, Bridget's crystal-blue eyes began sinking into gray sockets, her ribs protruded through her flesh, and she started walking around in a daze. At a tryout to place her in the regular or advanced group of tennis students, she failed to even get a racket on balls she would have drilled a month earlier. In words that humiliate me even more now than they did in 1979, I criticized her from the sidelines for "doggin' it."

Her pediatrician referred us to an endocrinologist, who admitted her to Children's Memorial Hospital in Chicago. Bridget began receiving insulin intravenously. Within a week, she looked and felt back to normal. The only difference was that her islet cells couldn't be revived or replaced. As any macho dad would, I fainted and collapsed on the floor as the doctor was telling us that the disease is chronic and incurable. (Dozens of people visited Bridget in the hospital, and the joke became to ask the nurses how Jim was doing.) I eventually managed to learn that the tip of one of Bridget's tiny fingers would now have to be stabbed with a stainless steel lancet three or four times a day, allowing a droplet of her blood to be smeared across a chem strip that measured the glucose in her bloodstream; this determined the correct dosage of regular and timed-release insulin to be injected into her butt or thigh. She would no longer be able to eat the same treats her classmates' parents had packed in their Big Bird lunch boxes or served at their birthday parties. But at least Bridget was released fairly healthy from Children's. At least she had come home at all.

Full disclosure: even if my daughter weren't ill, I would cheer on stem cell research with gusto. Because that's the kind of Mick I am, brother. I was raised Roman Catholic but have lived the last forty years as a secular humanist. People like me get branded atheists, heretics, ethical relativists, French, and much worse; more affectionate terms include freethinker, agnostic, existentialist, beatnik. I have faith that our bodies — brain waves and action, commerce and science, art and language and children — are pretty much all there is to us. Or all there is to me, anyway. Once my EEG line goes flat, it's gonna be all she wrote. In the meantime, my one and only life will be awful in several respects, awesome in others. Later or sooner, a crosstown bus or a Hummer will squash me like Wile E. Coyote. Either that or an organ or two will break down and I'll suffer, piss and moan not a little, then purchase the farm. I've already made the down payment.

As far as the suffering goes — when it starts, how long it drags out — I used to be confident that Western medicine was riding full speed, or almost that fast, to the rescue. I pictured a rangy Asian cowgirl in a lab coat and Stetson, a bandolier of specimen vials jangling against her modest cleavage as she clutches in one hand the reins of a galloping stallion, in the other a filament delicate enough to boink a few islet cells into a failing pancreas. Not that I expect to live forever — just an extra decade or so, with a little more spring in my step. More important, I want people like Bridget to get a fair shot at their biblical threescore and ten. Our Bible-totin' president, however, has stripped my infinitely resourceful cowgirl of her most promising protocols and forced her to ride sidesaddle on a stubborn Texas mule better equipped for wagon-train duty than galloping into the future.

Many of us who were willing to give Bush a chance, who voted for Al Gore but weren't terrified at the thought of a compassionate conservative in the Oval Office, now feel that the president isn't much less benighted than the Muslim fundamentalists he has engaged in an infinitely perilous clash of theocracies. Not that Bush is a dyed-in-the-wool Holy Roller himself. No, I believe he's a cold-blooded opportunist, able and willing to pander to our least-educated citizens — and that's much, much worse. After all, if, as Gallup recently reported, only 49 percent of Americans accept Darwin's theory of evolution, that number must hover near zero among certain blocs of red-state voters. What besides *shazam* can we say to these folks? But surely we can't let them hinder the momentum of biomedical research.

Not all conservative Christians attack science and reason while gorging on their fruits — laws, markets, medicine, weaponry, energy systems, computers — but the fanatical fringe surely does. Pretending to split the difference between these folks and the rest of us, the president's "compromise" effectively amounted to a ban on embryonic stem cell research. Announcing his decision from his ranch in Crawford, Texas, in August 2001, he claimed that "more than sixty genetically diverse stem cell lines already exist" and that the National Institutes of Health would be permitted to fund research on these existing lines only. Three years later, however, a mere nineteen of those lines have been made available to scientists, and most are genetically limited to people who tend to use in vitro clinics — the white, the infertile, the affluent. Not that there's anything wrong with these characteristics, but researchers will need hundreds — possibly thousands — of lines to provide genetic matches for the entire population. As the White House itself acknowledges in a fact sheet on the subject, "approximately 128 million Americans" stand to benefit from this research.

Fortunately, a movement to challenge or circumvent the Bush policy is now in full swing. On February 24, New Jersey governor James McGreevey submitted a budget that would make his state the first to fund embryonic stem cell research. In April the privately funded Harvard Stem Cell Institute opened with a mandate to fill the void left by the ban against federal money. On April 28, 206 members of the House — including 36 Republicans, more than a few pro-lifers among them — sent the president a letter urging him to loosen restrictions on embryonic stem cell research. On

June 4, the day before Ronald Reagan succumbed to Alzheimer's, a similarly bipartisan plea from fifty-eight senators landed in the president's in-box. And in November, Californians will vote on the Stem Cell Research and Cures Initiative, a hugely ambitious ballot measure that would allocate $3 billion in state money for research on stem cells, embryonic and otherwise. But so far, Bush hasn't blinked.

By supporting a bill proposed by Kansas senator Sam Brownback to send scientists engaged in embryonic stem cell research to prison and to outlaw treatments developed in other countries using such methods, the president has made this research downright unsavory, and he's done it on purpose. Even Dr. Kass regrets that young grad students won't go into stem-cell research as readily and that "maybe there's a certain chilling effect on the field as a whole."

Whatever is motivating President Bush, his policy has about it the stench of the witch doctor. It may be the most unenlightened position, with the most negative and far-reaching human consequences, ever taken by an American president.

In the 1980s, as Bridget's mother and I researched the possibilities for a cure, we were told to supervise Bridget's diet and insulin routine as closely as possible, and to teach her to maintain it herself. This would reduce the risk of complications ten or twenty years down the road. The better her control over her blood-sugar levels at each stage of her life, the healthier she'd be when a cure was discovered. It could even affect whether she'd be eligible for cutting-edge treatments.

During the eight or nine years after her diagnosis, our well-behaved little girl dutifully followed her regimens of shots, diet, and exercise. She almost never asked me, "How long do I have to take the shots, Dad? Do you know?" Having scrutinized every syllable of the Juvenile Diabetes Foundation literature and traveled to New York and St. Louis to interview researchers, I learned — or I chose to believe — that the disease would be conquered long before Bridget developed any serious complications. What I emphasized to her was that a cure would be found by the time she got her driver's license. "Just hold on until then," I would say with a hug, "and we'll have a ginormous double celebration."

At thirteen, Bridget was an A-minus student, a not-bad cello and piano and tennis player, and the starting shortstop for the Winnetka All-Stars, our town's traveling softball team. During the day-long tournaments on baked-clay diamonds in midwestern heat and humidity, she was usually the last player to run out of gas. We had to pack our Igloo with extra Gatorade and fruit and syringes, but by then we were used to that stuff. Bridget may have started sneaking the occasional candy bar or Pepsi with her teammates, but her overall health remained phenomenal. One mid-August Sunday, against the Deerfield Does, she snagged a line drive just behind second base, stepped on the bag to double off that runner, and ran down the runner advancing from first, completing an unassisted triple play.

Back at the Wyndham, the debate around the five-table pentagon has been framed, as it should be, in moral-philosophical terms. The tone remains cordial. Members forgo honorifics and refer to one another as Frank, Leon, Karen, "my friend." Yet the rock-bottom question persists: should "man in his hubris" or some other entity write the ground rules for biomedical research?

Chairman Kass has written, "In leading laboratories, academic and industrial, new creators are confidently amassing their powers and quietly honing their skills, while on the street their evangelists are zealously prophesying a posthuman future." A key notion for Kass is the Wisdom of Repugnance, also known as the Yuck Factor or, as I think of it, *Ewi*sdom. "Repugnance is the emotional expression of a deep wisdom," he writes, referring to such things as rape, murder, incest . . . and somatic cell nuclear transfer. "Shallow are the souls that have forgotten how to shudder." But the shudder test is hardly foolproof and can lead to a slew of false positives. When Dr. Zabdiel Boylston used inoculation to thwart Boston's smallpox epidemic in 1721, one clergyman thundered, "For a man to infect a family in the morning with smallpox and pray to God in the evening against the disease is a blasphemy," language that caused bombs to be thrown into the homes of Dr. Boylston's cohorts. Whereas hamstringing researchers racing to cure diabetes makes me shudder with rage and disgust.

As a Jewish M.D. and humanities professor, Kass has no trouble pronouncing words like *nuclear* or *vekhen lo' ye'aseh* (Hebrew for "such as ought not to be done"). He hardly fits the stereotype of the born-again zealot embodied by John Ashcroft, Dick Armey, and the like. That said, his views may surprise you. Kass's seven-hundred-page tome, *The Beginning of Wisdom: Reading Genesis,* performs what he calls an "unmediated reading" of the first book of the Bible. Nature is morally neutral, but humans should not be. The "new way," according to Kass, is really the old way, and can be summarized in a single word: *patriarchy.* Man rules in this world, and woman obeys — to her benefit. From the president's point man on ethics, we read that "a prolonged period of barrenness" before childbirth is God's way of "taming the dangerous female pride in her generative powers," and that marriage as an "institution of stable domestic arrangements for rearing the young depends on some form of man's rule over woman." Again, this is no adjunct lecturer at Bob Jones University, an easier target of condescension for radical centrists like me. Nope, this is the Addie Clark Harding Professor of the University of Chicago's Committee on Social Thought, that bastion of pricey Nobelitude.

I sit down with Kass during a break at the Wyndham. Trim and energetic at sixty-five, scholarly without seeming fussy, he has a genial spirit to go with a thorough command of this complex scientific material. Much as I disagree with him, I have little reason to doubt his good will. I begin by congratulating him for leading such nuanced and respectfully argued discussions for and against therapeutic cloning. I also tell him up front that, mainly because of my daughter, I want the spigot open much wider than he and the president do.

He politely objects to my premise. "It's been misrepresented that the council came out in opposition to embryonic stem cell research by a vote of ten to seven," he says. "It did not. It came out in favor of a ban on all cloning, including the cloning of embryos." Speaking, as always, in muscular sentences and pausing for paragraph breaks, he goes on. "Congress has not declared stem cell research illegal; it has said there should be no federal funding for research that involves the destruction of embryos." Pause. "The president found a way to fund embryonic stem cell research on the existing lines, which means he has liberalized the research opportunities."

Liberalized? Only if we're talking about the handful of lines he let slide in 2001 while banning federally funded work on the hundreds more lines researchers will need to succeed. But this is the schizoid position in which Kass and his boss have been in lockstep from the outset — or so I assume.

"Is it fair to say that the president appointed you chair of the council because he knew in advance you agreed with him about the proper limits of cell research?" I ask. "Or did he form his position after, or mainly because of, advice you have offered?"

"The truth is, I don't know," Kass responds. "I will say that I think it's improper to reveal conversations that we have had or to specu-late on the mind of the president. I think I know what brought me to his attention. It was not my views on stem cell research, about which I'd never written a word. I've been trying to stop human cloning for thirty-five years. The president wanted to hear a discus-sion of the ethical issues, to hear how one would lay out the various arguments. I think it's fair to say that I did not tell him what I thought he should do."

So it's not that the president wanted a council that would lend intellectual heft to whatever position wins red-state votes?

"Look," he says, holding my eye, "the president is pro-life. You've got to acknowledge it." He hooks a thumb back toward those of his colleagues still being interviewed. "But no previous bioethics coun-cil had anything like this diversity. It's not fair to say you've got a council that's opposed to embryonic stem cell research. You have some people on this council who'd be distressed to see lots of em-bryos destroyed before there was any proof of efficacy. I myself am very hopeful that over the next decade we will learn an enormous amount from the existing lines. The spigot is open."

"What about your opposition to in vitro fertilization?" I ask, re-ferring to the fact that in 1971 he was speaking against it aggres-sively, predicting that it would lead to deformed infants. A million or so normal IVF births down the road, he's been branded every-thing from a "bio-Luddite" and "false prophet of doom" to "a six-teenth-century sensibility to guide us through twenty-first-century conundrums."

"I was an early critic of IVF because I didn't know it was going to be safe for children," he says. "I thought IVF might eventually lead to cloning, which it might. I had a change of heart in 1978, just when Louise Brown was born. On the question of IVF for infertility, at least if it's shown to be safe, I changed my mind."

It takes a good man to admit that. Even so, I believe Dr. Kass is dead wrong once again. Given his way thirty years ago, we never would have discovered the glories of IVF; countless families wouldn't even exist. If Kass, Brownback, and the president succeed in making therapeutic cloning a criminal act, American scientists not only won't discover the first round of cures for spinal-cord damage, diabetes, Alzheimer's, Parkinson's, and heart disease, we'll never know how many other cures we missed out on. The tragedy will become exponential.

In the meantime, a thunderclap. On February 13, scientists in Seoul, South Korea, announced that they had succeeded in cloning human embryos and extracting stem cells from them. The team was led by Drs. Woo Suk Hwang and Shin Yong Moon, both of Seoul National University. Dr. Hwang emphasized that the research was subject to rigorous oversight by an ethics committee. It took place in test tubes and petri dishes; no embryo was, or could have been, implanted in a uterus. None of the sixteen volunteers who provided 242 unfertilized ova were paid. The project was funded by the government of South Korea, where reproductive cloning is illegal. The team's goal, Dr. Hwang stated forcefully, was not to clone human beings but to advance understanding of the causes and treatment of human disease.

Scientists and patients around the world hailed the results. A Chicago embryologist spoke for a lot of us by saying, "My reaction is, basically, wow." Dr. Kass's reaction came even before the Koreans' paper was published in *Science*. Speaking for the President's Council, he said, "The age of human cloning has apparently arrived: Today, cloned blastocysts for research; tomorrow, cloned blastocysts for baby-making. In my opinion, and that of the majority of the council, the only way to prevent this from happening here is for Congress to enact a comprehensive ban or moratorium on all human cloning." In much the same spirit, Carrie Gordon Earll of Focus on the Family branded the South Koreans' research "nothing short of cannibalism." But I'm here to tell you that *my* family — and, I assume, millions of others — was thrilled. Said Bridget, *"Finally!"*

The overarching fact that our president and the majority of his council seem unable to get their minds around is that human nature *evolves.* "Normal human" used to describe a four-foot-nine-inch club-wielding cretin draped in gore-spattered fur. It used to be "natural" for Tarzan, after impregnating Jane, to lope off in search of a new sperm receptacle, clubbing other men to death as he went, until — *blam!* And for his witch doctor to smear this wound with dung while grunting a few incantations. From the perspective of these early humans, *we* are what a Cro-Magnon Charles Krauthammer would fearfully denounce as "a class of superhumans," or the Addie Clark Harding caveman would hastily label "posthuman."

Glorious though it may be, evolution is also, as James Watson put it, "damn cruel," mainly because genetic mutations have introduced about fifteen hundred diseases into human DNA. What could we do about it? Not much until doctors like Rudolf Virchow (1821–1902) came along. Virchow was the German physician who finally convinced the medical establishment that the basic units of life were self-replicating cells. Building on François Raspail's axiom *Omnis cellula e cellula* — every cell, diseased or otherwise, originates from another cell — Virchow overturned the conventional wisdom that it was the entire body, or one of its "vapours," that became sick. Yet Virchow's most radiant brainchild may be that "medicine is a social science, and politics is nothing but medicine on a large scale." We still fall woefully short of this ideal, but as beneficent M.D.s like Hwang, Moon, William Mayo, Jonas Salk, and Paul Farmer have shown, fighting back against disease is the most humane thing we can do. The most human.

Whatever epoch we live in, we all have to face getting caught at the worst possible point on the curve of medical progress: my cowgirl's campfire is visible on the horizon, yet I am accorded the honor of being the very last hombre to succumb to Syndrome X. "Remember when people had heart attacks?" some lucky duck in 2050 may guffaw, clutching her chest in mock agony. "I mean, can you *imagine?*" This fortunate woman would be exactly as human as any victim of plague or polio, and a lot more human, in my view, than the cretin with the club. However long it may seem to us now, her life span will still seem to *her* the way Nabokov and Beckett imagined it in the middle of the twentieth century: as a brief crack of light between two infinities of darkness.

*

Speaking of infinities of darkness, adolescence is the first prolonged test for people suffering with type 1 diabetes. Physical rou-
tines readily followed by obedient ten-year-olds suddenly become a
series of temptations to rebellion — against parents and authority
in general, against the unrelenting regimens themselves. The flood
of new hormones disrupting skin tone, academic performance,
and household peace also wreak havoc on a diabetic's cardiovascu-
lar system. Psychological anxiety goes thermonuclear, which sets
off more fuming rebellion. My doctors and parents say I can never
smoke cigarettes? I'll retreble my efforts to buy them at 7-Eleven.
And what harm can come from skipping my shot on homecom-
ing night? When Bridget's doctor revealed that her sugar control
wasn't nearly as tight as it should be, she started skipping blood
tests as well. Her mantra became "Why should I struggle to take
care of myself if I'm gonna die young anyway?"

In the meantime, I stepped up my visits to researchers, who by
the early 1990s were working on ways to keep the islet cells of cows
and pigs from being rejected by humans. I studied reports, wrote
letters to congressmen, talked to more doctors. I imagined myself
going blind; I found myself volunteering to "God" to deal with this
fate instead of Bridget. My firstborn child, my talented and beautiful daughter, was being ravaged from the inside out by a rapist who
was taking his time, and there was nothing I could do about it.

I began a novel about riding a bicycle from Chicago to Alaska.
Narrated in the voice of a young woman with diabetes, it was my attempt to empathize as vividly as I could with the disease-ridden
angst my daughter faced every day. *Going to the Sun* turned into a
road trip connecting two love stories, and the character of Penny
Culligan is an amalgam of myself, my sister Ellen, and Bridget. But
Penny's diabetes is the plutonium rod — potent, relentless, explo-
sive — fueling each strand of the narrative. The ending is designed
to evoke both the hope and desperation the young woman feels as
she makes her way into adulthood.

The longer you have this disease, the more severe its complications become. It's a hassle from day one, to be sure, but after fifteen
or twenty years your kidneys begin to break down and your retinopathy becomes more severe. Bridget's had diabetes for twenty-
five years now. She's frightened, exhausted, and angry. She's also
determined to overcome her long actuarial odds and live some-

thing resembling a normal life. But your self-esteem takes big hits when you have a chronic disease. Your skin becomes sallow and spongy from all the punctures; you also get to worry about whether you can get pregnant, carry a baby to term, then survive long enough to see your child enter kindergarten. "Why should I have to listen to the history of your cold," Bridget sometimes wants to know, "or how tough your meeting was? At least your freakin' pancreas works!"

If a cure isn't developed in the next few years or so, Bridget will become more and more susceptible to a heart attack or stroke, however diligently she takes care of herself. She may still go blind, and her kidneys might fail. As her circulatory system gets ravaged further, the dainty feet she used to lace into white size-5½ softball spikes may need to be amputated.

What scares me the most, of course, is the possibility that Bridget's disease will take its course just before the cure comes on line. Which is why I so heartily agree with the council's minority. American society, it writes in *Human Cloning and Human Dignity,* has "an obligation to heal the sick and relieve their suffering." The minority accords "no special moral status to the early-stage cloned embryo," because it has no capacity for consciousness in any form. Even more to the point, "the *potential* to become something (or someone) is hardly the same as *being* something (or someone)."

As to whether somatic cell nuclear transfer would lead to cloned children or human-animal hybrids — the slippery-slope argument — Harvard's Michael Sandel proposes that the research should proceed "subject to regulations that embody the moral restraint appropriate to the mystery of the first stirrings of human life." Specifically: strict licensing criteria for labs, laws against commodifying ova or sperm, and measures to keep private firms from monopolizing access to cell lines. The minority concludes that it's "perfectly possible to treat a blastocyst as a clump of cells usable for lifesaving research, while prohibiting any such use of a later-stage embryo or fetus." We shouldn't outlaw all cloning, in other words, just because its therapeutic applications *could* be misused. Even Michael Jackson's face doesn't get plastic surgeons arrested.

Finally, because of advances in somatic-cell nuclear transfer, *any* human cell could theoretically become a person if it were doctored aggressively enough in a lab. "If mere potentiality to develop into a human being is enough to make something morally human," the minority continues, "then every human cell has a special or inviolable moral status, a view that is patently absurd."

In addition to Sandel, the enlightened minority includes Janet Rowley, the Blum-Riese Distinguished Service Professor at the Pritzker School of Medicine at the University of Chicago. Dr. Rowley is blunt in her opposition: "Our ignorance is profound; the potential for important medical advances is very great. Congress should lift the ban and establish a broadly constituted regulatory board, *now.*"

A few hours after my chat with the chairman, I sit down with Dr. Rowley. "Dr. Kass makes the case that the council hasn't done anything to inhibit embryonic cell research," I begin. "He claims Bush has *opened* the spigot."

"Well, I would disagree," says Dr. Rowley, an elegant, hazel-eyed woman with undyed silver-brown hair. "For many people in the majority, a fertilized ovum is a human being, and therefore taking that single cell and doing anything with it is murder. Others of us believe that while a fertilized ovum has the potential to become a human being, if this potential human being could in fact lead us to something that could save the lives of many human beings, then we think it's a matter of competing goods. Helping many, many individuals is justification for taking a single cell or an early multiple-cell organism and using it to benefit more individuals."

Before I can hug her or put my fist in the air, she continues: "Now, to a purist, of course, that's immaterial. It's 'man in his hubris intervening,' not nature or God or whatever entity you want to invoke."

"Yet God is never mentioned explicitly . . ."

"No," Rowley says. "They would generally cite moral and philosophical reasons. But in our early conversations, it was brought up that in Jewish and Islamic law, a developing embryo didn't become human for forty days — well within the limits of the time frame in which embryos would be used for therapeutic research."

Think back to that human-development continuum in your high school biology textbook. Sperm meets ovum on the left, full-term fetus to the right. No reasonable person would let researchers destroy anything recognizably human, just as no one objects to putting blood under an electron microscope. The points of contention fall somewhere between a blastocyst and the forty-day-old embryo of religious tradition. Exactly where we draw the line, and whether we make a distinction between what happens in a petri dish and what *could* happen in a uterus *but doesn't*, is determined by either spiritual/faith-based beliefs or rational/scientific principles.

In a democratic society, then, who gets to draw it? Why not George W. Bush, for example, with the backing of his eminent council? Well, one reason is that the person officially charged with helping Dr. Kass push the majority's agenda through Congress is executive director Dean Clancy, who spent eight years on the staff of Dick Armey. Not only is Clancy opposed to embryonic stem cell research, he virulently opposes public schools and federal taxes, which makes him what most folks would call a fanatic. Such appointments extend the Bush-Cheney pattern of loading any dice that get tossed down on the policy table. In December 2002, for example, Bush appointed Dr. W. David Hager to an FDA committee on reproductive drugs. Hager is not only ferociously opposed to therapeutic cloning but has been accused of refusing to prescribe contraceptives to unmarried women; to women suffering from premenstrual syndrome, he counsels prayer and Bible reading. (You heard me.) In February, a group of sixty scientists — including twenty Nobel laureates — issued a statement that this administration repeatedly censors reports written by its own scientists, stacks its advisory councils, and disbands those offering unwanted advice. "Other administrations have, on occasion, engaged in such practices," the scientists wrote, "but not so systematically nor on so wide a front." Bush's response to such criticism? In late February, he and Dr. Kass replaced two members of the minority — Elizabeth Blackburn and William F. May — with three new members who all oppose therapeutic cloning. Dr. Kass's fair, diverse council is now tilted thirteen to five.

The majority of a different president's council — John Kerry's, for example, or the one Al Gore would have appointed — could make an equally strong case in favor of therapeutic cloning. At least as many alternate chairmen with bring-us-to-our-knees credentials — Janet Rowley, for example, or Douglas Melton, co-director of the Harvard Stem Cell Institute — stand ready to argue that position. Most of our scrutiny, then, should focus on the person who makes the appointments.

In January 2002, the president greeted members of his council in the Roosevelt Room of the White House, asking them to be mindful of "the notion that life is — you know, that there is a Creator." It's precisely this willful confusion of realms, this thumbing of his nose at the sworn constitutional duty to keep church and state separate, that keeps folks like me from stomaching the president but makes millions of others just crazy about the guy.

Isn't it inevitable, then, that sectarian dogma will tip the balance on every tough issue? Not really. Plenty of conservative Republicans, such as Trent Lott, Orrin Hatch, and Nancy Reagan, support embryonic stem cell research. Senator Hatch has said, "I just cannot equate a child living in a womb, with moving toes and fingers and a beating heart, with an embryo about to be taken from the freezer and which will be lawfully discarded if we don't use it."

Sooner of later, of course, a Bush will need cell-replacement therapy, and he or she may have to fly to California, New Jersey, or maybe even Seoul. Much worse, the necessary treatment may not be ready yet *anywhere*. But here's another thing I have faith in: once the cures are available, those who opposed therapeutic cloning in 2004 will damn well find a way to get themselves and their families treated. It's not even hard to imagine them elbowing their way to the front of the line at the clinic. "Oh, yeah?" says my daughter. "Over my dead body."

American medicine has long been guided by men and women of serious learning, not religiously correct politicians. We seem to understand that when science gets trumped by sectarianism, more bad things happen than good. The framers of our Constitution deliberately omitted *God* from its language, assigning supreme power to "We the People." They wanted to insulate us from holy wars, crusades, and oxymorons like "creation science."

Jimmy Carter, our first born-again president, faithfully kept religion and policy separate, and the intelligence of his heart gets more and more plain every year. In the early 1960s, moderate Democrat John Kennedy went out of his way to avoid giving even the slightest impression that his Catholicism might override his duties as chief executive. John Kerry, another Yankee Catholic, gives every indication he would do the same thing.

Back in July 2001, Kerry wrote a letter urging Bush to fully fund embryonic stem cell research. As Kerry said later, "Compassionate conservatism could have meant lifesaving treatments for those suffering from Parkinson's and Alzheimer's disease; instead it appears to be using words of compassion to mask efforts to keep a campaign promise to conservatives . . . If, as he says, the president believes that stem-cell research may have lifesaving potential for millions, he should give scientists the tools to explore it rather than have the government impose burdensome restrictions." Kerry then cosponsored (with Republican Arlen Specter) a Senate bill to support embryonic stem cell research. The bill stalled in committee, but now, on the campaign trail, Kerry pounds away on this issue: "The medical discoveries that will come from stem cell research are crucial next steps in humanity's uphill climb . . . If we pursue the limitless potential of our science — and trust that we can use it wisely — we will save millions of lives and earn the gratitude of future generations."

In the meantime, here's how a wartime Republican balanced civic and spiritual responsibilities back in 1862, while thinking about the Emancipation Proclamation: "I am approached with the most opposite opinions and advice, and that by religious men who are equally certain that they represent the divine will . . . I hope it will not be irreverent for me to say that if it is probable that God would reveal his will to others, on a point so connected with my duty, it might be supposed that he would reveal it directly to me . . . These are not, however, the days of miracles, and I suppose it will be granted that I am not to expect a direct revelation. I must study the plain physical facts of the case, ascertain what is possible, and learn what appears to be wise and right." And he did.

SHERWIN B. NULAND

Getting in Nature's Way

FROM *The New York Review of Books*

SCIENTIFICALLY ADVANCED NATIONS, most notably the United States, seem on the verge of a new situation in which the traditional goals of doctors and others concerned with health care will be radically altered. The changes will be the result of increased understanding of the basic molecular mechanisms shared by all living things and a widened ability to devise technological methods by which those processes may be manipulated. More than a few observers of the biomedical scene believe that we are about to enter an age in which the improvement of human bodies and minds will become a primary goal in research and clinical treatment. Some of these observers are hopeful; some point to the possibilities of inherent and unforeseeable danger.

Until little more than a century ago, the only aims of medical care were the cure of disease and the relief of human suffering. But the definition of "human suffering" has gradually changed. We now find ourselves faced with the reality that it is no longer sufficient to prevent or treat sickness of the body or mind, that physicians are expected to address increasing attention — and society's dollars — to the millions who are dissatisfied with what nature and their own DNA have given them. Whether for rhinoplasty, botox injections, or a prescription for sex hormones, thousands of men and women daily make their way to doctors' offices, intent on improving themselves. Not sick in any usual definition of the word, such discontented people would like to be better than they are, better than merely well. Even "better" may not be enough. That is why Sheila and David Rothman have called their cautionary new book *The Pursuit of Perfection*.

If the pronouncements of some futurists are to be believed, enhancements of human appearance and function will soon be so effective and commonplace that many will wonder in coming years why some critics scoffed when Gregory Stock, director of the Program on Medicine, Technology, and Society at UCLA, called his book *Redesigning Humans: Our Inevitable Genetic Future*. The title of Stock's opening chapter was "The Last Human," by which he meant those few remaining whose bodies and minds have been formed by nature and nurture alone. As they age to more than double the biblical three score years and ten, the contented beneficiaries of the coming technologies may look back with scorn at the bioethicists and others who criticized William Haseltine, the biotech entrepreneur and CEO of Human Genome Sciences Incorporated, when he proclaimed to a *New York Times* reporter, "I believe our generation is the first to be able to map a possible route to individual immortality."

These are the outcomes envisioned by the pioneers who believe a biogenetic gold rush will soon take place. But every new technology carries the possibility of introducing unacceptable risk as well, which is why the subtitle of the Rothmans' book is *The Promise and Perils of Medical Enhancement*. The Rothmans deal not with the future, but with such recently popular treatments as hormone replacement and cosmetic surgery. Showing how these have been evaluated may help to avoid mistakes in administering treatments. We should profit, say the Rothmans, from the errors of excessive certainty, popular zeal, and self-interest that have accompanied some recent innovations. And we would profit also from a frank assessment of the sometimes deliberate and sometimes inadvertent collusion between researchers, pharmaceutical companies, and practicing physicians that has enlarged the market for enhancement.

Sheila and David Rothman are social historians of medicine whose writings have brought attention to the ways in which communal, economic, and political forces combine to influence decisions that many of us naively thought were shaped only by the needs of patients and the progress of science. Two of their books have become minor classics among bioethicists because of their careful documentation and thoughtful, if partisan, analysis of events that led to significant changes in patient care as well as attitudes toward the medical profession. In their 1984 book *The Willowbrook Wars: A Decade of Struggle for Social Justice,* they described the lawsuits brought by parents against the Willowbrook State School, a home for retarded children and adults in New York City with notoriously wretched living conditions and an abusive staff. Like the inmates' parents, the Rothmans believed the inmates could cope with life on the outside more effectively than the state authorities imagined. They showed that reforms would never have taken place if parents, public organizations, and politicians had not become involved. In their view, improvements in health care policy will take place only when the public demands them. It is not from within the medical profession that reform is accomplished, but from without.

In 1991, David Rothman published *Strangers at the Bedside: A History of How Law and Bioethics Transformed Medical Decision Making,* chronicling the rapid evolution, primarily between 1966 and 1976, of the doctor–patient relationship, which rendered it less paternalistic and gave patients more autonomy as it brought new participants into the process of health care. Much of the change came about as the result of the social upheavals of the late 1960s, which gave rise to increased demands for self-determination and hastened the establishment of bioethics as an academic discipline. The most forceful public expression of the principles of medical self-determination occurred with the *Roe v. Wade* decision of 1973, prohibiting state laws that restrict a woman's right to abortion during the first trimester of pregnancy. By then, medical treatment was being influenced not only by increasingly insistent patients but by the courts, ethicists, the writings of social scientists and legal scholars, and an ever more knowledgeable public demanding to know how decisions concerning medical research and clinical care were actually being made.

Rothman argued that increasing pressures to carry out clinical research after World War II conflicted with medicine's historic commitment to the individual patient. The size of the research establishment grew enormously in the two decades following the war, as the National Institutes of Health were rapidly enlarged and the federal government poured huge amounts of money into university health centers, most of which was directed toward clinical and laboratory investigation. Academic promotion came to depend far less on teaching and patient care and far more on the publication of papers in scholarly journals. The result was a weakening of the bond between doctor and patient, filtering down from the medical school faculties to the doctors they trained. The situation was aggravated when new technology raised ethical questions about prolonging life, sometimes pitting doctors against patients. In 1976 there was public clamor over the case of Karen Ann Quinlan, a comatose young woman whose doctors refused to honor her parents' request to remove her breathing tube. They kept her alive by artificial means for thirteen months while her parents sued to be allowed to take her off the respirator.

Other developments combined to make a stranger of the doctor at the very bedside where he had once been the patient's advocate and source of reassurance: the increasingly impersonal atmosphere in hospitals, the introduction of widespread third-party payment and managed care organizations, the creation of the new field of bioethics, and the growing involvement of the courts. The doctor's influence was eroded by those newer strangers, among them ethicists, lawyers, insurers, and administrators of managed care. Though Rothman did not spell it out, each new participant in medical care brought about a further weakening of the personal responsibility of the doctor — the responsibility that was once the patient's best guarantee of the concerned care that is the single most important element in successful clinical treatment. The Rothmans recognize, moreover, in view of the complexities of modern medicine, the increasing fragmentation of the profession into specialized units. A patient may be referred to several different doctors for a single complaint.

In turning their attention to the problem of medical enhancement, the Rothmans consider yet another burden on physicians whose prime motivation has been the thoughtful care of the sick. The fact is that enhancement, whether through female hormones or liposuction, has a variable record, not only having often failed in its intention, but too frequently having exposed patients to unanticipated hazards — such as infection, increased risk of cancer, and even death — or to complications that might have been predicted if many members of the public were not so quick to accept innovation. Scientists, pharmaceutical houses, popular magazines, advertising agencies, and even clinicians themselves are carried along by the excitement of research advances and the eagerness of potential consumers, as well as by the prospect of making money. An enthusiasm takes hold that sweeps caution before it.

These influences are, in the Rothmans' words, "reinforced by a culture that prizes individual perfection and peak performance," and they are concerned about its implications for the future:

> The system, however, is out of balance, for no part of it has a stake in emphasizing or even communicating the dangers that are almost certain to accompany the innovation . . . The record strongly suggests that technologies will emerge slowly and haltingly, some delivering benefits, others inflicting serious harm. Consumers will be compelled, personally and collectively, to make a series of exquisite choices, with very little data to guide them . . . Healthy adults will have to calculate how much risk they are willing to accept in order to try to optimize a trait. Is it wise to undergo an intervention that promises to dramatically increase life span and disregard the risk that it might cause fatal disease and shorten life span?

It is this record of "slow and halting" innovation that the Rothmans address in their thoroughly documented and readable book. "What science creates medicine rapidly dispenses," they warn, and this uncritical acceptance by both physician and consumer is precisely the problem.

The Rothmans begin their narrative with a long and complex account of female menopause and its discontents. Not long after the discovery of hormones around the turn of the twentieth century, medical scientists began to view these chemical compounds as the basic determinants of physiological functioning. The very word — hormone — is the clue not only to their action but to the hopes for their manipulation as well, being derived from the Greek verb *hormaein*, meaning "to excite" or "set in motion." Extracts of animal testicles had already been thought (falsely, it was later shown) to restore youth and potency to aging men, and the search was soon on to discover the vital male force they contained, as well as an analogous female regenerative compound in the ovary.

When the female hormones called estrogens were identified, it seemed natural for physicians and women alike to promote their use in restoring the femininity lost following menopause. At first estrogens were prescribed to maintain youthful skin, hair, and out-

look on life, and later to combat the distressing symptoms of menopause and its aftermath, such as hot flashes, insomnia, and loss of
bone density. In time, statistical studies involving follow-up of many
patients seemed to suggest that properly used hormone replacement therapy (HRT) would also decrease the incidence of heart
disease, breast and genital tract cancer, and Alzheimer's dementia.

But no matter its other presumptive advantages, the emphasis was
always on HRT's benefits in reversing, or at least holding off, the
aging process. Though the women who took hormones may initially have been attracted by their effect on menopausal symptoms,
most of those who continued treatment were more interested in
how they looked, the texture of their skin, and the revitalization of
their energies. To many of them and to their doctors as well, aging
was a disease, unnecessary and henceforth treatable. If taking a few
pills every day could enhance the quality of life without any dangers, as the press and television proclaimed with the wide agree-
ment of the medical profession, why not try them out?

Some foresighted gynecologists warned that estrogen's potential
for encouraging tissue growth might promote the development of
cancer, but calls for specific studies of that possibility went un-
heeded. And then in 1971, it became apparent that women whose
mothers had taken the synthesized estrogen compound diethyl stil-
bestrol (DES) during the first trimester of pregnancy had a distinctly higher risk of developing adenocarcinoma of the vagina, a
relatively rare malignancy. A few years later, two studies were published describing a several-fold increase in endometrial cancer, a
cancer of the uterus, among women who had been on HRT. The
medical profession's response was to change the treatment, adding
progesterone, a steroid hormone, and emphasizing the preventive
aspects of the hormones, especially the well-documented improvement in bone density for those women at risk for osteoporosis. It is
estimated that more than a third of postmenopausal American
women were having hormonal therapy by the end of the 1990s.
Meanwhile, further investigations had failed to confirm that it lowered the incidence of Alzheimer's disease.

Finally in July 2002, a study by the U.S. government–funded
Women's Health Initiative comparing 65,000 women on hormones
to 100,000 controls was stopped because it had become obvious

that those on HRT were having more coronary events, strokes, blood clots in the lung, and invasive breast cancers. Though they had fewer colorectal cancers and hip fractures than the control group, the decision about HRT was expressed as follows in an editorial in the *Journal of the American Medical Association:* "Do not use estrogen/progestin to prevent chronic disease."

Of course, there is still a place for HRT in the lives of many women who, having considered the risks and benefits with their doctor, may find it useful to take the hormones for their own distinctive physical situation. Some, for example, have such distressing menopausal symptoms that they feel justified in accepting the added possibility of danger; so do women who by family history or for some other reason are at such a high risk of developing osteoporosis that the benefits would seem to outweigh any other considerations. But at least the word has spread that such treatments do not guarantee a life of youthfulness and that decisions about chemicals whose primary purpose is enhancement must be made individually by each patient and her doctor. The real question for physicians and consumers, though, as posed by the Rothmans, is whether the kind of careful testing by the Women's Health Initiative that limited the use of HRT will be applied to other treatments. "Will they now use HRT as the model for guiding their use of plastic surgery and liposuction? Will they use the story of estrogen as a template for evaluating future genetic enhancements?"

To respond to the first question, there is no evidence that the demand for liposuction has slowed down. Over four hundred thousand such procedures to remove fatty tissue were performed in 2001 (of which 20 percent were done on men), and the number grows ever larger. This in spite of statistics indicating that some eighty of those patients died as a result of the operation, a mortality rate exceeding that for adult hernia repair by a factor of almost seven. Hundreds of thousands of Americans are subjecting themselves to a possibility of postoperative complications and death that surgeons would find unacceptable for any other elective procedure. It is ironic that the doctors who choose to perform an operation that is solely cosmetic are willing to accept mortality and complication rates significantly higher than those who restrict their interventions to those required for the treatment of disease. Perhaps this says something about the standards observed by cosmetic surgeons. Yet we can expect, as the authors write, that liposuction will continue to "go forward without significant attention to risks."

Ironically, the earliest of all hormonal manipulations has never established itself as either popular or particularly useful. Though hormone treatment for testosterone-deficient young men is an established medical intervention, the notion that it will rejuvenate the elderly or improve their sexual functioning has found little support in unbiased studies. Following decades of initial optimism, there seems far less interest in it on the part of either doctors or patients. One nowadays finds only infrequent glowing testimonials to testosterone of the sort that were common in women's magazines to boost the sales of HRT. There was a burst of advertising for testosterone by drug companies in the decades following World War II, but this has significantly decreased in recent years.

The reasons for the falling off of enthusiasm for testosterone supplements go beyond the fact that they have not been shown to accomplish the purpose for which they were initially touted. Because there is no male equivalent to the dramatic changes of menopause, symptoms requiring urgent amelioration — insomnia, hot flashes, weight gain, and so on — do not occur, and the changes in appearance set in gradually. Moreover, middle-aged men tend not to be as intent on maintaining youthfulness, whether in appearance or physiology, as women of the same age. When they get a medical checkup, they are more likely to have it done by an internist than are women, who frequently use their gynecologist as a general practitioner. A visit to a specialist in urology, who would be far more likely to recommend testosterone therapy, only occurs if the actual symptoms overcome the man's greater reluctance to seek medical attention.

And then there is the question of cancer. Like estrogen's effect on breast tissue, testosterone can stimulate cellular proliferation in the prostate. Studies have not been extensive enough to prove an association beyond doubt, but contrary to the experience with estrogen, physicians are hesitant to prescribe a therapy with such potential when there is no proven benefit. The National Cancer Institute and the National Institute on Aging, concerned that steadily increasing use of testosterone compounds might result in a greatly heightened incidence of prostate cancer, recently asked for advice from the Institute of Medicine of the National Academy of Sciences on how to convince the public of the hormone's lack of benefit and its potential dangers. The academy has suggested small clinical trials to provide evidence of risk. Should the results of such investigations prove unrevealing, large, long-term studies could then be undertaken.

And yet some physicians continue to prescribe male hormones. Though the FDA prohibits the marketing of testosterone as an antiaging therapy, there are enough hints in the advertising of such products and in occasional articles appearing in the popular press that many physicians will prescribe it for selected patients. Moreover, the hormone is a staple — along with other kinds of hormones, fetal cells, and numerous antioxidants — of the hundred or more so-called rejuvenation clinics that have sprung up throughout the country. The result is that anyone who wants testosterone without his doctor's knowledge can get it, whether by answering an advertisement or by going on-line to a Web site that provides the names of doctors who treat "testosterone deficiency."

Among the rationalizations for giving testosterone to older men is that their natural levels of the hormone wane with age: it would seem logical to use replacement therapy even though it seems to provide no benefits. Something of the same logic has been used to justify injections of growth hormone, whose blood levels were also found to decline with the passing of years. Growth hormone treatments, often administered to children with a deficiency of the hormone, had been a source of contention for decades, not only because a number of cases of the neurologically crippling Creutzfeldt-Jakob disease were found among children to whom it was given, but also because of the vexing question of which children should receive it. Should a short but not hormone-deficient child be treated? How short is short? Where does treatment end and enhancement begin? Because the use of growth hormone in some children resulted in stronger bones, increased muscle strength, and reduced body fat, researchers thought it might reverse some of the common problems of aging, namely osteoporosis, muscle weakness, and the accumulation of fatty tissue. Not only have the results been ambiguous, but significant side effects have occurred, such as joint pain, numbness, and swelling of the legs. In addition, experimental work indicated that mice injected with growth hormone did not live as long as those without it.

None of this has deterred the rejuvenation clinics and many physicians from prescribing growth hormone. Though not approved by the FDA to treat the symptoms of old age, there is no law against such use, and so-called off-label prescribing for older men is common. "All of which," conclude the Rothmans, "helps explain why enhancement technologies, whatever their putative benefits or demonstrated risks, will have significant space in our future."

And, they add, "there is no holding back the enterprise." Research will go forward, and there will be great pressure for its clinical application:

> As this history of enhancement has demonstrated time and again, routine methods of oversight will not be adequate, nor will the advice of individual physicians or professional medical societies or government regulators. What is required is an intimate understanding of the nature of the research and the reliability of the results. Only with this information at hand will consumers be able to calculate potential risks and benefits to know whether to join the line outside the doctor's office, or to demur.

This is wise advice, and the Rothmans have built a powerful case for it. But it is far more easily given than taken. In the long run knowledge guarantees neither wisdom nor sound judgment. Seeking out every available fact about some method of enhancement — or any other medical intervention, for that matter — does not give the perspective that can come only from professionally trained authorities with experience in the distinctive variety of critical thinking that is called clinical judgment. The fact that the clinical judgment of physicians has been woefully inadequate in the situations so well described by the Rothmans does not mean that it should be discarded.

What it does mean is that some of the strangers previously mentioned do have a place at the bedside. Decisions that in decades past were considered strictly clinical must now be recognized as having a moral, an ethical, a philosophical, and a legal aspect, and even a bearing on public policy. Ideally, the therapeutic implications of every coming medical advance should be scrutinized with these perspectives in mind. When that becomes the norm, soci-

ety and individual patients — and the Rothmans — will need no
longer fear that practitioners, medical societies, or government
will abdicate their responsibility. What might be proposed for such
scrutiny is a variation on today's bioethics committees, in the best
of which physicians and nurses with scientific or clinical expertise
join with ethicists, lawyers, the clergy, and community representatives to recommend a course of action that arises from the consensus of the group. The makeup of such committees might vary with
the therapy being evaluated and its possible implications. While no
system of oversight can be flawless, such committees may not only
discover and publicly communicate problems that might arise with
new technology but also bring attention to matters that should be
considered by more specialized advisers.

The evidence that such a state of affairs may be attainable comes
from American experience with end-of-life care. Since the Karen
Ann Quinlan case in 1976, there have been many changes in the
way decisions are made during every phase of the process of dying.
The current wide availability of hospice care is an example of that,
as are the frequent use of such legal strategies as durable power of
attorney or the appointment of a health care proxy, vast improvements in palliative care, not to mention its being established as a
distinct medical specialty, and the greatly increased involvement of
families and patients.

In making such changes, the medical community has by and
large responded with heightened sensitivity to the advice of philosophers, bioethicists, and even lawyers. My own impression is that
clinicians are far more understanding, empathetic, and skilled in
dealing with dying patients than they were a quarter-century ago.
Pointing this out is not to imply that the demands of patients and
families do not have a decisive effect, but we know that the impetus
for change could not have been accomplished without the involvement of the experts and advisers I have mentioned.

I wonder whether it is true, as the Rothmans claim, that "there is
no holding back the enterprise." It is just possible — now for the
first time in the history of modern science — that the moment has
finally come when society might reconsider whether the curiosity
and enthusiasm of scientists alone should determine the direction
of research into certain technologies. As biomedical investigation
moves into the forms of enhancement that will affect personality,

intelligence, memory, organic structure, and longevity, perhaps we ought to make use of our experience with those strangers at the bedsides, and bid them visit not only the clinic but the laboratory too.

To calculate what the Rothmans call "potential risks and benefits" is praiseworthy, but in order to do that one must have better knowledge of those risks. The misadventures that these writers portray in their important book prove that we enhance ourselves at our own peril, and much of that peril is yet to be discovered.

To accomplish the feats of genetic improvement predicted with such assurance by Gregory Stock and William Haseltine is to forget the admonition of Francis Bacon, who was, after all, the father of the scientific method: "Nature, to be commanded, must be obeyed." Two centuries earlier, Michel de Montaigne had warned of the dangers of doing otherwise when he pointed out that we should not get in nature's way, because "she knows her business better than we do." Long before the Rothmans, such philosophers were putting us on notice.

SHERWIN B. NULAND

The Man or the Moment?

FROM *The American Scholar*

IN A RECENT LETTER to the editor of the *New York Review of Books*, the eminent physicist and author Freeman Dyson was taken to task for his essay on a new biography of Isaac Newton. "Dyson unfortunately shows how little versed he is in scholarship on Newton," wrote the correspondent, a faculty member of the Division of Humanities and Social Sciences at the California Institute of Technology. The basis for this denunciation was Dyson's apparent failure to realize that "historical research requires deep understanding of very different technical issues of long-gone science, together with substantial knowledge of social and cultural circumstances of the period." Though the writer went on to accuse Dyson of specific factual errors, he seemed most exercised by what he considered his quarry's misunderstanding of the era in which Newton worked.

The letter made me chuckle, and I'm certain that it had a similar effect on many other observers of the decades-long conflict between the social historians of science and those who might be termed the technical historians. In the unlikely event that the characteristically unflappable Dyson felt any heat from his antagonist's discontent, he might enjoy knowing that the temperature is even higher among scholars who study the history of medicine.

Until some forty years ago, the technical historians of medicine dominated the field. They were almost all physicians, and hardly any of them had even the most minimal training in the formal methods of historical scholarship. Their interest was in the landmark achievements of clinical and laboratory medicine, and in the lives of the men responsible for them. To their self-appointed task they brought an expertise born of intensive medical training and extensive patient care. They focused only peripherally on wider historical currents. These doctors were, in the strictest sense of the word, amateurs. When they slipped away from their consulting rooms to dusty library stacks — or when they elected to forgo Caribbean vacations in favor of research trips to old hospitals in far-off countries — they did it for love of the grand tradition of their forebears, to which they considered themselves heir.

Things began to change in the 1950s, as the field of medical history became increasingly professionalized. After that time, the academic degree held by authors of scholarly articles was as likely to be Ph.D. as M.D., and before long the former predominated. Within twenty years, the majority of participants in the annual meeting of the American Association for the History of Medicine had earned doctorates in history. Among the most prominent were men and women who held both titles and had abandoned patient care to devote most or all of their time to historical studies.

With this shift came a change in focus from the technical and personal to the societal and cultural. No longer was it sufficient to investigate the exploration and the explorer; the intellectual atmosphere of an entire era was now scrutinized. Medical science was seen as the product less of individual genius than of the Zeitgeist. Discoverers were shaped by their times, not the reverse.

The bedside doctors welcomed these new insights, but they were not happy to watch their perspectives being shoved aside by people with little or no clinical background. Though both medical doctors and social historians have something valid to say, the latter are currently in the ascendancy. The old-style physician-historians have been routed, and in the process, much has been lost.

By 1980, Leonard Wilson of the University of Minnesota, a Ph.D. and the editor of the *Journal of the History of Medicine and Allied Sciences,* warned of the consequences of a history grounded completely in cultural causes and dominated by scholars who "see little of the laboratory and less of the clinic." He said of such scholars: "They tend to neglect questions of clinical medicine, of the biology of disease, and of science, even when such questions had a direct bearing on the particular historical subject with which they are concerned. The result is incomplete and sometimes severely distorted history . . . If such social history be considered medical history, . . . it is medical history without medicine." And one might add that it is medical history without the colorful characters who made it.

Beyond a doubt, there exists a cultural inevitability to scientific discovery. The sun would have been recognized as the center of our solar system whether or not Copernicus had lived, and probably soon after he published his monumental *De Revolutionibus Orbium Coelestium* in 1543; the debunking of the phlogiston theory was in the cards, and Lavoisier merely speeded the process; the discovery of the structure of DNA would have taken only a few more months had not Watson and Crick outsprinted everyone else to the finish line. In each case, the times were ready; the ambient culture and the state of contemporary science virtually ensured that these advances would occur, and fairly soon.

At least in medicine, the precedents were in place for every discovery by the time it was presented. Even the transcendent contributions of Harvey and Pasteur would have been made had those two brilliant men never been born, though they would have taken place somewhat later. But they would have been made in a different way, usually as the result of a different process — because part of the process is the distinctive personality of the discoverer.

Far more often than most social historians are willing to admit, a discovery made at a particular time and place — and the form in which it is brought to the community of medical thinkers — is unique to the person who is responsible for it. Not only that, but it not uncommonly arises from the idiosyncrasies of that one individual, and may even be the expression of his or her personality, background, or personal situation. Similarly, when a contribution is not readily accepted, the failure can often be ascribed as much to the way the innovator has come to it and brought it to attention as to a cultural milieu not yet ready to embrace it. Of all that is "incomplete" and "severely distorted," and of all that is lost by the social historian's downplaying of individual effort in favor of surrounding influences, the one missing factor that most diminishes the ultimate narrative is the unique personality of the contributor, and the ways in which it plays into the process of discovery and the overall cavalcade of history.

Regardless of the surrounding culture, some scientists are aggressive while others are mild-mannered; some are resentful of authority while others do precisely what their teachers expect; some are intolerant of delay while others achieve their ends through patient persistence. These characteristics profoundly affect the nature and timing of landmarks in medicine. When Thomas Carlyle wrote that "history is the essence of innumerable biographies," he was referring specifically to the role of inimitable individuality in shaping the events of our world. Even today, when discoveries are often team efforts, it is ultimately the single observer or experimenter who must initiate the process of his or her own contribution to science. The fact that the same contribution would eventually have been made by someone else does not in the least vitiate the force of that truth.

Examples abound. Throughout the 2,500-year history of Western scientific medicine, progress has repeatedly been spurred or slowed by the personal behavior of an individual. Most prominent in this regard during the classical period was Galen of Pergamon, the second-century physician whose many public demonstrations of animal experimentation energized the doctors of his time and explained physiological phenomena previously obscured by a hodgepodge of spurious theories. So powerful was the effect of his research, his performances, and the many dozens of books he left to posterity that his influence towered over the meager efforts of his successors. But Galen was a vain, fiercely competitive self-promoter, driven as much by the search for eternal glory as by the search for knowledge. Summoning all the authority he had earned from his scientific and clinical contributions, he declared that his teachings were to be regarded as the unchanging gospel of medicine. There was no point, he taught his eager acolytes, in attempting to seek new information about health, disease, or the structure and function of the human body; he had created a complete and sufficient system of medicine.

Such was the forcefulness of Galen's personality, as expressed both during his life and through the enduring influence of his writings, that his teachings were blindly followed for nearly a millennium and a half. Medicine stagnated in his honor until the sixteenth century. At that point, along came Andreas Vesalius, an ambitious and endlessly curious young man as contentious as his Greek predecessor, among whose most striking personality characteristics was a love-hate attitude toward authority figures that culminated again and again in angry conflict. He stood up to his teachers, first at the University of Paris, then at Bologna and Padua. Finally, he took on Galen himself, denouncing the ancient master for the more than two hundred errors he had made in his anatomical descriptions and exposing the reason: all of Galen's work was done on monkeys and dogs. In 1543, at the age of twenty-eight, Vesalius wrote a monumental book, *De Humani Corporis Fabrica,* that founded the scientific study of anatomy and established for medicine the principle that progress can be made only by taking tiny steps — and by challenging authority. Henceforth, medical innovators would abandon the old reliance on the conception of grand theories into which observations must be uncomfortably pigeonholed.

In my own specialty, surgery, there are abundant examples of men whose personalities left their mark on the course of medical history, Zeitgeist or no Zeitgeist. In 1837, a young Hungarian obstetrician named Ignac Semmelweis, in a moment of inspired brilliance, discovered the reason that almost 20 percent of the obstetric patients in virtually all of the major European hospitals were dying of childbed fever: the obstetricians were not washing their hands after performing autopsies on the pus-ridden bodies of the women who had died of the same disease within the previous twenty-four hours. Without a microscope, and long before germs had been recognized as the agents of disease, Semmelweis intuited that "invisible organic matter" on the hands of the doctors was being conveyed into the genital tracts of women in labor, consigning them to an anguished death. But he was a self-righteous and combative man, and he alienated his superiors and most of his colleagues by accusing them of remorselessly murdering women when they would not accept his theory without experimental proof. After a halfhearted attempt to provide such evidence using a few rabbits, he refused to do further laboratory work, contemptuously declaring the truth of his assertion to be so self-evident that no additional studies were needed. He saw every attack on his doctrine as an attack on himself. Semmelweis would die in a Vienna mental asylum, beaten to death by orderlies trying to restrain him. His great discovery was forgotten, and the promulgation of the germ theory, which would have occurred around 1840 had he been less bullheaded, was delayed until 1867.

When the theory was finally brought forth in that year by the gentle, supremely patient Quaker surgeon Joseph Lister, the notion of microscopic organisms causing disease seemed so outlandish — and even foolish — to the physicians of the time that it found little general acceptance. It was the quiet persistence and good-natured equanimity of Lister — along with his continuing experiments, his demonstrations, his writings, and his willingness to travel from hospital to hospital to disseminate his beliefs — that finally won the day, though that day was delayed for some two decades.

Two more examples from widely separated historical periods illustrate the significance of the role sometimes played by an innovator's personality, each presented independently of the era and aura in which he worked. The first is the story of an introverted, asthenic little Breton physician, René Théophile-Hyacinthe Laennec, who stood five feet three inches tall and, at the age of thirty-five, had never spent an hour alone in the company of a woman who was not a relative or a servant. One day in 1816, on hospital rounds, Laennec faced the terrifying obligation of putting his ear — in the manner of the time — directly against the chest of an intimidatingly pretty young woman in order to hear the transmitted sounds of her lung disease. The pathologically shy little man backed off and hurried home. En route, he chanced upon some boys playing a game familiar to him, in which the scratchings of a pin on one end of a long rod of wood were interpreted by a lad listening at the other end. Struck with inspiration, he hurried back to the hospital, rolled up a sheaf of papers into a cylinder, placed it under the left breast of the amused girl — and in that historic moment invented the instrument he called *le baton,* soon to be refined into what we now know as the stethoscope. Laennec carved his own *batons,* which could be purchased for three francs as a sort of supplement to the thirteen-franc book he would write three years later, describing the many uses of his new diagnostic tool.

The second example is culled from the annals of American medicine. William Halsted, a brilliant and dashing young New York surgeon known for his speed and technical derring-do in the operating room — as well as for his high living — was among the first experimenters with cocaine after it was shown, in 1884, to induce local anesthesia when injected into skin or muscle. Having no idea of the drug's dangers, Halsted used himself as an experimental subject and soon became our nation's inaugural cocaine addict. Following a long period of attempted recovery, he secured an appointment as the first professor of surgery at the new Johns Hopkins Medical School shortly after it opened in Baltimore in 1893. Halsted emerged from the darkness of his cocaine-saturated period as a very different man from the fearless risk taker of his early career: a methodically slow, meticulous operator whose painstaking methods would be emulated by the many surgeons he trained. The so-called Halstedian technique in time spread throughout the country and made possible a new "surgery of safety," as it became known, and the introduction of many innovative operations that required great gentleness and minute attention to detail. This withdrawn, asocial professor is remembered today as the father of American surgery. His methods of dealing with tissues and organs, essential if further progress was to be made, would not have been introduced for years or even decades had it not been for the personality alteration that resulted from their inventor's cocaine addiction.

Though my own fascination with medical history lies most with the people who have made it, I would never claim that this perspective is always the most effective one. Several factors — social, cultural, technological, and personal — can be explored separately, and for the sake of analysis they may be treated as independent variables. Nevertheless, the process of discovery arises from all of them working together, each in its proper proportion. Some variables may be more consequential than others in any given case, but they are all crucial to the evolution of the bit of progress being studied. The punishment for devaluing the significance of any of them is the writing of bad history.

JEFFREY M. O'BRIEN

To Hell and Back

FROM *Wired*

"IAN DROWNED." In a cave 4,500 feet beneath Oaxaca, Mexico, with an underground waterfall roaring in the background, Bill Stone didn't hear those words as much as he sensed them. He knew that the odds of survival at close to a mile deep aren't great for even the strongest cavers, a chiseled superset of explorers who routinely haul 180-pound loads down 65-foot vertical shafts, live for weeks underground, and dive through bone-chilling water. Mapping the last uncharted land on earth is a dangerous proposition. One slip, one momentary lapse of concentration, and you die.

At the time of his death, Ian Rolland had been using Stone's latest invention — the Mk-IV, a device now referred to as a rebreather, which scrubs exhaled air and recirculates it, greatly extending the amount of time a diver can stay underwater. Still experimental when Rolland strapped it on, the Mk-IV had the potential to push cave exploration to ever greater depths. Each unit had triple redundancies and was debugged in countless simulations, but the Oaxaca expedition was the real thing. When Rolland's body was found at the far end of a flooded, 1,400-foot-long cavern, the natural assumption was that the innovative equipment had somehow failed.

There was only one way to find out: someone would have to retrieve the body and the equipment, in all a 300-pound load. With his team leery of the new technology and exhausted after a dozen days underground, Stone knew it was up to him. He'd have to traverse the sump himself and carry Rolland back through a stalactite-riddled black pool of near-freezing water, breathing the recycled air being pumped into his lungs by his own contraption. And hope to hell that whatever happened to Rolland didn't likewise happen to him.

It might be harsh to call Stone's house in Gaithersburg, Maryland, the worst on the block, but it does lack curb appeal. A muddy, aging Toyota Tacoma sits in the driveway alongside an untended lawn. Stone answers the door wearing a cap that bears the name of his bar band, the Terminal Syphons. "I thought you'd be the pizza guy," he says, a set of braces showing under a Fuller-brush mustache and sunken cheeks.

Over the course of two summer evenings, we sit at a card table and discuss invention, risk, NASA, and Stone's latest quest: to develop an intelligent autonomous hydrobot. With a doctorate in structural engineering and eleven patents to his credit, Stone is the archetypal modern-day explorer, a multidisciplinary maverick constantly inventing tools in the name of discovery lust. Any Ph.D.'s résumé will feature a few — maybe even a few dozen — peer-reviewed articles. Stone has published more than 220 papers on topics ranging from seismic resistance to spacecraft development.

These days, he's absorbed by two projects. The first: Stone is the chief architect of a next-generation ladar (laser radar) system at the National Institute of Standards and Technology, helping craft a beer can–sized guidance system for unmanned military vehicles. And then there's DepthX, an audacious NASA-funded project formally known as the Deep Phreatic Thermal Explorer. It is a robot that will map an ocean six miles beneath the frozen surface of Jupiter's moon Europa and sniff out microbiological life. A team of researchers — roboticists, astrobiologists, and electromechanical engineers — from six universities are working on the project. But it's Stone's baby.

Even though DepthX is about robotically exploring a moon 400 million miles away, it's a direct result of Stone's life as this world's preeminent cave explorer. He originally conceived of the hydrobot as a way to map a Mexican cave, Sistema Zacatón. A series of limestone sinkholes, Zacatón is a fascinating mystery: toxic hydrothermal vents on the cave floor make it impossible to traverse much below 50 meters. What if, Stone wondered, he could send a robot to map the areas that humans are unable to reach? That inkling, combined with his engineering know-how and a $5 million grant from NASA, became DepthX.

The project fits a pattern of spontaneous invention for Stone. Faced with the impossible (or impassable), he always seems to devise a way around it. Over the past thirty-three years, he has spent 353 days below 1,500 feet. He's invented breathing regulators, diver propulsion devices, and 3-D mapping tools. When it became clear that lugging dozens of scuba tanks into the depths was hindering his ability to plumb farther, he designed a closed-circuit rebreather to eliminate the tank farms. The rebreather works by recycling expired air, removing poisonous carbon dioxide and looping clean oxygen back to the diver. The original model had twin gas processors, four onboard computers, and enough air to last forty-eight hours. He tested the prototype by spending twenty-four continuous hours underwater. Rebreathers have since become standard issue in both cave exploration and deepwater diving.

"Many people think all I do is some random engineering work in between caving expeditions. It's been far more deliberate than that," Stone says. There's a fusion, he adds, between "an explorer and the tools you use to explore the frontier."

Stone's obsession has come with costs. His austere lifestyle — no TV, no stereo, little in the way of furniture — testifies to the scant thought he gives to the usual pleasantries of everyday life. Like many of his kind, he's lost a marriage to exploration. But even the most accomplished explorers marvel at Stone's commitment to an incredibly dangerous discipline. "Bill takes risks far beyond those taken by astronauts," says the former New Mexico senator and *Apollo 17* astronaut Harrison Schmitt, "with much less external support."

When he's underground, Stone borders on possessed. Team members insist that when life is on the line, there's nobody more reliable. But he can also be oblivious to the limitations of those around him. "You have to have a keen sense of your own limits when working with Bill. He will push you as hard as he can," says Kenny Broad, an anthropologist and professor at the University of Miami who was on the 1994 Oaxaca expedition. "A lot of people get offended by that. But the way Bill sees it, we're all adults. It's up to you to self-regulate."

Stone definitely has trouble finding his own off switch. Between ladar and DepthX, he's been working eighteen-hour days, and he's dog tired. But he perks up when the discussion turns to risk. He has no patience for a society that is obsessed with safety but overlooks those willing to jeopardize their own lives. "There are plenty of people on Earth. It's not like the human race is going to disappear if a few people don't come back," he says coldly. "Exploration is dangerous. So NASA has lost seventeen people. Here's a list of sixteen of my close friends dead with no state funerals, no schools named after them. These people have been to unknown territory, where every step forward is one step deeper into what might be the world's deepest cave. It's a holy mission."

The Oaxaca expedition started out normally enough. After testing the rebreathers off the Florida coast, Stone's team of forty-five explorers headed for the cave known as Sistema Huautla. Within a few days they had rigged the cave with two miles of rope and were prepared to break through to the other side of San Agustín Sump, a low point in the cave not unlike the j-trap on a bathroom sink — but full of silt and stalactites instead of toothpaste and bobby pins. More than a quarter-mile long, San Agustín had proven impassable in previous expeditions; traditional scuba tanks don't last long enough at such depths. With a rebreather, Kenny Broad cracked the sump on his first try, reaching a sand bar on the other side. He returned to Camp Five, and Ian Rolland excitedly set off to scoop some booty — a caver's term for discovering new terrain. "Don't call out the cavalry unless I've been gone for six hours," he told Broad.

Most often divers set out in pairs. In cave exploration, having a buddy only increases the complexity — kicking up silt, reducing visibility — and does little to increase safety. If one diver gets into trouble, a rescue attempt will likely mean the loss of two lives rather than one. So when Rolland left, he knew he was on his own.

Broad became antsy around the third hour. The fact that Rolland was a diabetic didn't help. He had been diving for years, managing his affliction with candy bars. What if he had forgotten his candy bar this time? After six hours, Broad rallied a search team. A few hours later, he set out for Rolland. When he cracked through to the other side, he found his partner — floating in a few feet of water.

After Broad returned, Stone set off, determined to bring the body back and find out what had gone wrong. He strapped a rebreather on his back, dove into the black water, and disappeared into the sump.

An hour later, he arrived at the other side. The area around the body showed no signs of struggle. Rolland appeared to have died in his sleep, his regulator dangling at his side. Stone clipped Rolland's body and rebreather to his chest, his friend's lifeless head hanging before Stone's face. Stone eased back into the water and descended, clutching the guideline that Broad had laid. Stone adjusted Rolland's buoyancy device to keep from sinking, but he had to be careful. Too much air and both would instantly shoot to the roof of the sump.

Which is just what happened. Stone overinflated the vest, causing him to surge upward and lose his grip on the rope. He was off the line in a flash, trapped in complete darkness between a dead body and the roof of the sump. In minutes, hypothermia would set in.

At 484 million miles from the sun, with an average surface temperature of −260 degrees Fahrenheit, Europa is no place for humans. And yet many planetary scientists agree that it represents the best hope for finding some form of life elsewhere in our solar system. That's because when *Voyager* flew through the Jupiter system in 1979, it found indications of a liquid ocean — under six miles of ice.

NASA has a three-stage plan to search that ocean. First comes the lander, followed by a nuclear-powered cryobot to melt through the ice. The third stage depends on Stone's DepthX. It will descend thirty miles, propel itself around the ocean while mapping the area and sniffing microscopic lifeforms, then send the collected data over an acoustic modem to the cryobot. If DepthX works, it will represent not only a huge boost to NASA's quest to find extraterrestrial life, but also a significant advancement in robotic intelligence.

DepthX will get its first test in 2007 at the source of its inspiration, Sistema Zacatón, which happens to be a neat analogue for Europa. Clearly, the flooded limestone sinkhole isn't as cold as Europa's ocean, but it can be incredibly toxic, extends to unknown depths, and has proven impervious to humans. (The world's premier cave diver, Sheck Exley — who happened to be a friend of Stone's — died in Zacatón in 1994.) "We know that life is there within the top forty meters or so," Stone says. "Below that, no one knows what's there."

Stone's bot will map Zacatón with a sonar-based technique known as simultaneous localization and mapping, which Stone used in 1998 to create the first 3-D underwater cave map. The bot will collect samples to 500 meters down, employing a search method Stone calls "hierarchical detection and discrimination." Simply put, it categorizes all lifeforms in familiar environments and then excludes that information in the extended search of the toxic environment below. It's kind of like using the minus sign to improve a Google search — you want results about Prince, but nothing about Prince Charles. In effect, the bot is learning as it scans, eliminating all known data to improve a search for the unknown. "The question is, Can you train a robot to look for the things that would guide you to the unusual bits of life?" Stone wonders. If the answer is yes, the hydrobot could be on its way to Europa by 2015, according to Stone's estimates. But getting a robot into space is not enough for Stone. He wants to be there himself.

In the exploration game, astronauts sit at the top of the heap. But Stone has never been able to count himself among them — though not for lack of trying. In 1989, he and fifty-nine other big brains made it to the semifinal round of NASA's astronaut selection process. He had been working toward the moment his entire life, crafting his education to make him more appealing to the selection committee. After repeated applications, he obtained a pilot's license (like Neil Armstrong), and then a Ph.D. (like Buzz Aldrin). Now NASA wanted nineteen people. One in three finalists would make the cut. Stone thought he had a shot — until the thin envelope arrived. "They determined that I was too independent for teamwork," he says, his face souring. "I said, well, that may be how you see it in your psychological analysis, but I look at how many times I've been on expeditions where I've done nothing but haul other people's equipment. Even when I'm leading a project, my mantra is, Let those who want to be out front be out front."

Fifteen years later, Stone is still smarting. He considers NASA to be a bureaucracy populated by politicians who are afraid to do anything great — and he's determined not to let the agency keep him from space. He has a $5 billion plan to get to the moon within seven years. He wants to bring together industry and government on a mission to set up a commercial mining outpost on the moon, where he and a crew will live for two years and establish "a moon-Earth economy" by mining hydrogen from the Shackleton crater at the lunar south pole. The hydrogen will be used to create water and rocket fuel, which will get Stone and his team back to Earth — they'll be going there with only enough fuel for a one-way trip.

It all sounds laughably sci-fi. But Stone talks more passionately about this project than anything else. And his plan does have sympathizers. *Apollo 17*'s Schmitt advocates drilling the lunar surface, though he proposes mining helium. Paul Spudis, a planetary scientist at the Johns Hopkins Applied Physics Lab and a member of the President's Commission on Moon, Mars, and Beyond, applauds Stone's intent. "He's trying to short-circuit a lot of the pointless negativity about space exploration and get something going," says Spudis. "I'm of the same vein. We need to be more entrepreneurial and risk-taking."

Where will Stone ever get $5 billion? He's courting oil companies interested in tapping into a new source of fuel. Exxon Mobil could pay for the whole project with a single quarter's profits. NASA thinks Stone is nuts. Chris McKay, a planetary scientist at NASA Ames Research Center who is working on DepthX, calls him a "brilliant engineer" but the proposal "daft."

But that's not stopping Stone. "Well, I'm more dangerous than they think," he says. "With a motivated core of techies backed by guerrillas, you can do some pretty amazing stuff on a small budget. If we can just find the money, we're out of here. We'll wave goodbye from low-Earth orbit."

The natural human response in an extremely stressful situation is to hyperventilate — take in more oxygen than otherwise necessary. On land, overbreathing causes lightheadedness and maybe some finger tingling and can be rectified by breathing into a paper bag. Underwater, it's deadly. A hyperventilating diver will run through his air supply at ten times the rate of a relaxed diver. With a scuba tank, Stone easily could have sucked through all of his oxygen while thrashing against the sump ceiling. But no amount of aggravated gasping will deplete a rebreather's capacity. Rather than releasing precious oxygen in bubbles, a rebreather scrubs out carbon dioxide and recycles the rest of a diver's air. Stone may have been pinned against jagged limestone 4,500 feet below Mexico, lugging 300 pounds of cargo and staring into the eyes of his dead friend. But at least he wasn't going to run out of air.

He deflated Rolland's buoyancy vest to trigger a free fall. Legs and arms extended, he groped for the 2-millimeter-diameter line in a 40-foot-wide sump. Stone snagged the line with a lucky reach, composed himself, and resumed his dive. But halfway back, he let too much air out of Rolland's vest and began to sink uncontrollably. Stone lost the line again and punched at the buoyancy vest inflator, but couldn't reverse course. The half-dead tandem crashed to the sump floor with a thud, kicking up a mound of silt. Stone waited at the bottom and once again caught his breath. The water cleared, and slowly he ascended to the safety guideline, one more time.

Three grueling hours after Stone had set out, he returned, spent, with Rolland's stiffening body. Crew members covered Rolland's head with wetsuit hoods to keep it from decomposing (or to keep from looking at it), and over the next five days they hauled the body 4,500 feet to the surface. Then Stone faced another problem. After their seventeen days underground, a freak hurricane came whipping off the coast, dumping 10 inches into the cave, raising the water table by 20 feet, and trapping four team members at 3,000 feet down. Stone was aboveground at the time, waiting helplessly for the water to subside. Once the team surfaced days later, just about everyone bailed. True to form, Stone continued on, joined only by the most inexperienced member of his team, Barbara am Ende — also Stone's girlfriend. They would press on for eighteen more days, mapping another 3.3 kilometers of uncharted land — and naming a vast expanse after Ian Rolland. The final push established Sistema Huautla as the eighth-deepest cave in the world, but the couple never did reach the end. Stone believes that one day Sistema Huautla will prove to be the deepest on Earth.

The data recorder on Rolland's rebreather showed that the device functioned perfectly on his final dive. Doctors concluded that Rolland simply had a hypoglycemic blackout. The way Stone sees it, his friend had set out without a candy bar, became excited at discovering untouched land, and pushed his body beyond its limits.

Of the sixteen friends Stone has lost to cave exploration, he's recovered seven bodies himself. He says that having a friend die never gets easier. But rather than mourn Rolland's death back in the United States, he pressed on — because if he hadn't, the trip to the depths of Sistema Huautla would have been nothing more than an adventure tale. And Stone's not an adventurer. He's an explorer. The difference comes down to one simple word: information. "If you don't come home with data, you've done nothing," he says. "It's a stunt. It's a story. You've accomplished nothing."

IAN PARKER

The X Prize

FROM *The New Yorker*

ABOUT SIX YEARS AGO, not long after Burt Rutan, an aircraft designer, had begun to think seriously of building a plane that could leave Earth's atmosphere, he woke up at his desert home, a few miles outside the town of Mojave, California, and said, "I've got an idea." As Rutan's wife, Tonya, recently recalled, "We had been dead asleep — it was three in the morning. He ran to the bathroom with a sketchpad, came out waving a sheet of paper, and said, 'I know how to configure the spaceship! A shuttlecock.'" Her husband, she said, was "real close to my face, looking right in my eyes." Burt asked Tonya to think of the way a badminton shuttlecock falls: how it always falls in the same position, and falls slowly, its feathers causing drag. That would be a smart way, he said, to return a small spaceship safely to Earth. The couple went back to sleep, and Burt, who often starts the day by singing James Brown's "I Feel Good," woke in high spirits.

I was eating lunch with Tonya, a slightly built and pretty woman in her early forties, in the Voyager restaurant, which is attached to Mojave's airport. "See those mountains? We're cut off from the world," she said, nodding toward the nearby Tehachapi range. She said that she would prefer to live in a more cosmopolitan place, adding, "Fortunately, we get to travel a lot." Her hometown is a line of gas stations and fast food restaurants along a hot, flat desert valley floor a hundred or so miles north of Los Angeles. A new bypass has taken business away from the town; the local Wendy's closed last spring. A resident described Mojave as the kind of place where, if you were driving late at night across country, "your wife would look around and wouldn't let you stop."

But in Mojave the sky is almost always clear. At night the Milky Way is milky, and it rains only about a dozen days a year. Despite an insistent northwesterly wind that keeps windmill turbines turning on the surrounding mountain ridges, this is a good place to fly. An airport was built in Mojave in 1935; during World War II, the marines took it over. It became a quiet civilian airport in 1961; eleven years later, it was reorganized as the East Kern Airport District, and local farmers occasionally leased space between the runways to dry fruit. It's a frontier-spirit place, perfect for training test pilots or keeping a privately owned MIG fighter plane.

The view from the restaurant was of two hundred or so commercial jets sitting in the dry air, waiting for an international economic upturn; the airlines use Mojave for long-term parking. The only movement was that of a man riding an undersized bicycle up and down the airport apron. But a videotape running in a continuous silent loop on a television monitor on the wall pointed to activity inside the hangars and workshops on the airport's grounds: privately funded research into space travel. The film showed a spindly-winged white plane waiting on the runway, with a smaller, stubbier plane slung beneath it, like a missile. They took off; the film then showed a vertical vapor trail streaking way overhead, a thin white line. After this, the smaller plane returned, and Burt Rutan, waiting on the tarmac in jeans and a mustard-colored polo shirt, hugged the pilot, who had just become an astronaut. This preliminary flight, which took place in June, will be repeated on September 29 — this time with ballast equal to the weight of two passengers. Space tourism is about to receive its first official test.

When Dick Rutan, Burt's older brother by five years, crashed his model aircraft as a child, Burt would fix them and make them his own. Burt, who is sixty-one years old, spent much of his childhood, in the central California town of Dinuba, building spectacular, award-winning airplanes. Dick went on to join the air force. Burt flew solo on his sixteenth birthday, took a degree in aeronautical engineering at California Polytechnic State University, in San Luis Obispo, and then, in his twenties, worked as a civilian flight-test engineer at Edwards Air Force Base, close to Mojave. (In one of the audiocassettes he sent home in place of letters during this time, he said, "I really want to make something of myself in aviation . . . I want to make it before I get too old. I want to get there quick.") He

married in 1962, had two children, and divorced not long afterward. He later said, jokingly, that he saw a choice between his wife
and the plane he was then building, and "it was an easy decision to
make."

Rutan moved to Mojave in 1974. By that time, forty-eight Americans had been in space, and twelve had walked on the moon; Pan
Am had placed more than ninety thousand names on a waiting list
for future scheduled moon landings, and President Nixon had announced a program to build reusable shuttles that would have the
effect of "routinizing" space travel. Rutan, then thirty-one, was confident that before he died he would fly into space.

He went into the hobbyist business, designing avant-garde airplanes and selling the plans (at about $130 each) to amateur en-
thusiasts with large suburban garages. The planes tended to be ca-
nards, which have two sets of wings — one small set in front of the
other — making them almost impossible to stall. They were designed to be built from light, strong, composite materials — layers
of fiberglass and epoxy, for example — rather than wood or metal.
With the success of his awkwardly named VariEze ("very easy") and
Long-EZ aircraft, Rutan became a revered figure in the "home-
build" community. "Rutan is God," one hobbyist wrote.

In 1982 Rutan formed a company, Scaled Composites, that produced single prototype aircraft for commercial and military clients.
Four years later, he became known internationally when he designed *Voyager,* an aircraft with a wingspan of 111 feet, which was
built to fly around the world on a single tank of fuel without stopping — something that had never been done before. *Voyager'*s pilot
was Dick Rutan, who had moved to Mojave after retiring from the
air force; he had flown more than three hundred combat missions
in Vietnam. While Burt was intense and self-contained, occupied
by his private world of aeronautical invention, Dick was social and
self-confident. ("I'm the world's greatest pilot," Dick told me, when
I met him at the Denny's in Mojave. "I'm the right stuff. I am!")
The brothers frequently fought. Dick Rutan still sounds peeved, if
also amused, when he recalls Burt's reluctance to install radar in
Voyager. "He thought it was just some froufrou thing that I wanted."
Burt was more attuned to what a plane needed to fly than to what a
pilot needed to survive the flight. "Burt doesn't always understand
the human element in design," Dick said. *Voyager* was difficult to fly

and came close to killing him a number of times — first on takeoff, when its fuel-filled wings bent so far as to touch the ground — but in December 1986 it made a precarious 25,000-mile flight, in nine days. It is now displayed at the Smithsonian National Air and Space Museum, in Washington, D.C. Ronald Reagan presented the Rutan brothers with the Presidential Citizen's Medal.

By the time Rutan reached his fiftieth birthday, in 1993, the prospects for everyday manned space flight and a station on the moon — not to mention a manned voyage to Mars — were no closer than they had been twenty years before. The future had stalled; the space shuttle — sold to the country as an exercise in thrift — had turned out to be expensive and fatally unsafe.

Spurred by grievance and by some refinement of the feeling that inspires middle-aged men to buy Ferraris, Rutan began to make sketches of a vehicle that he hoped would be capable of reaching 100 kilometers, or just over 62 miles, above sea level — before quickly dropping back, like a flying fish, pulled by gravity. (In the mid-1950s, the World Air Sports Federation had defined this altitude as the beginning of space.) Rutan was not imagining an orbital craft, which would need to reach far greater speeds and would require many times the level of investment; rather, he was aiming to reproduce the parabolic suborbital flight made by America's first astronaut, Alan Shepard, who spent fifteen minutes above Earth's atmosphere in a *Mercury* capsule in May 1961. Rutan saw an opportunity to become the person who opened up space. "It became very clear that if the little guy didn't do it, it wasn't going to be done at all," he said.

An alternative space community — predominantly libertarian in spirit, and sometimes Vulcan in its wardrobe — had been pressing since the late 1970s for an end to NASA's historical monopoly on American space flight. That case was reinforced by the destruction of the *Challenger* shuttle in 1986. Even those who had once been inspired by NASA began to see the agency's early achievements as disguised disappointments. "The American space program was, for all practical purposes, an attempt to show the Russians that we could do Communism better than they could," Jeff Greason, the president of XCOR, a Mojave-based rocket-plane company, said recently. "I'm not sorry we went to the moon, but there's very little doubt in my mind that had *Apollo* never existed we would be further along today than we are." In pursuing grandiose projects, he said, NASA overlooked more practical approaches to space travel.

By the mid-1990s, more than a hundred civilian satellites had been put into orbit, and the idea of private rockets launching new satellites into space began to seem a realistic business proposition. At about the same time, the wilder idea of entrepreneurial manned space travel started to migrate out of science fiction. Some of those in the generation of cheated Americans — who had regarded the space race as a private invitation to the moon, then seen the party canceled — began to acquire great personal fortunes, allowing the alternative space rhetoric to be backed by capital. As Elon Musk, who sold PayPal to eBay for $1.5 billion dollars in 2002, and who now runs SpaceX, a rocket company near the Los Angeles International Airport, recently described it, "You had to have some kind of preboom to supply the capital to get the rocket boom going, and that only happened with personal computers and the Internet."

In 1996, Peter Diamandis, a St. Louis entrepreneur and space enthusiast, founded the X Prize, which was designed to induce an age of space tourism. Diamandis modeled his prize on the $25,000 Orteig Prize, which was won by Charles Lindbergh for the first transatlantic flight in 1927. The X Prize (in the aeronautical tradition, the letter "X" indicates "exploration" and "experimental") promised $10 million to the first privately financed team to take three people to the threshold of space and back twice in two weeks. The second trip would prove the spaceship's reusability.

Around this time, Burt Rutan met Paul Allen, the cofounder of Microsoft. Today Allen is ranked by *Forbes* as the third-wealthiest man in the world. (He recently spent part of his $21-billion fortune to found a Science Fiction Museum in Seattle that contains, among other exhibits, the captain's chair from the set of *Star Trek*.) Rutan told Allen of his plans for a rocket-powered plane and showed him some sketches. Allen was impressed, and they agreed to speak again.

A rocket is a bomb that explodes neatly, and to deliver someone into suborbital space is neither easy nor safe; fourteen people have died on space-bound rockets since 1960. But Rutan saw the greater challenge as bringing his spaceship back to Earth. As a spaceship reenters the atmosphere and begins to encounter air of increasing density, it must maintain a precise trajectory, or it risks breaking apart. At first, Rutan pictured a capsule that returned under a parachute, as in the *Apollo* and *Soyuz* programs. But to an aeronautical expert, this method was unappealing, and Rutan began to envision his spacecraft flying back to Earth, not falling. By 1999 Rutan's thoughts had turned to a concept close to that of the X-15 — the black, winged, experimental American rocket plane of the 1950s and '60s that was launched at 45,000 feet from a modified bomber plane rather than starting its flight from the ground. This high-altitude takeoff allowed the X-15 to carry 50 percent less fuel. From 1959 to 1968, when the program ended, the X-15 took eight pilots above 50 miles, which is the air force's designated start of space. In the opinion of many experts, the X-15 represented a future of elegant, low-cost, reusable spacecraft that was sidelined by America's investment in *Apollo,* which in today's terms amounted to about $100 billion.

The X-15 showed that it was possible to land a rocket plane on the ground. But a computer system was often used to fly the plane at the precise angle to reenter the atmosphere safely. Rutan did not want the weight, the expense, and the complication of that kind of system. (His brother had, after all, barely won the argument about the need for radar on *Voyager.*) It was at this stage that Rutan thought of badminton — and of the idea of a shuttlecock and what he calls "feathered" wings. A spaceship could have the wings needed to land on the ground, but they could be hinged and rotate, like oversized air flaps: this would cause drag, the way the feathers on a shuttlecock do, and would set the craft at a stable, belly-first angle on reentry.

In 2001 Rutan signed a contract with Paul Allen, agreeing to a budget that has been estimated at about $20 million. Rutan was not initially inspired by the X Prize. Neither was Allen, according to Rutan: "If you see how rich he is, Paul Allen wouldn't do it to try to win ten million." But the prize was there — at least until the end of 2004, when a policy that underwrote the contest would expire — and it would have been foolish to ignore it. So Rutan's designs included seating for three — a pilot and two passengers — as the X Prize required. "I had confidence," he said. "I just marched along at the rate I felt I could do it, not really worried about the competition. And, fortunately, there hasn't been a lot of competition."

*

By this summer, twenty-seven teams had registered for what is now known, because of sponsorship, as the Ansari X Prize. Nineteen of the teams were from the United States, and the rest were from Argentina, Canada, Israel, Romania, Russia, and Great Britain. Many teams had Web sites showing handsome vehicles far above Earth: the sky black behind them, the horizon curved, the oceans a lovely blue. Only a few of the competitors had been able to take their adventures further than these hopeful illustrations.

While the Orteig Prize had marked a clear aeronautical milestone, the X Prize, if awarded, would primarily celebrate a moment of economic confidence. In terms of space exploration, Yuri Gagarin's Earth orbit in 1961 was a greater achievement than any X Prize suborbital flight; and the prize's requirement of reusability had been met by the space shuttle in the early 1980s. The X Prize wasn't only about getting people in and out of space; it was about a new way of spending unusual quantities of private money. It championed extravagance; and in doing so, it demanded unlikely relationships. It called for nongovernmental finance: engineers had to find angel investors, or investors had to find engineers.

In Texas, John Carmack, who amassed a fortune writing video games like Doom and Quake, began developing Black Armadillo, a cone-nosed cylindrical spaceship that, at journey's end, would turn itself around, nose up, then gently settle to Earth using downward thrust. And this summer a Canadian team, da Vinci, that had long struggled to finance a project designed to launch a rocket at 71,000 feet from beneath a hot-air balloon, found funding from an Internet casino known for sponsoring promotional streakers at sporting events. Brian Feeney, da Vinci's founder, has agreed in return to play on-line casino games on a laptop computer in space.

While a dozen or so teams have devised potentially plausible technological plans, far fewer have had the funds to follow up on them. The level of these financial difficulties is suggested by a recent press release from the young team of Space Transport Corporation, in Forks, Washington: "The latest proud S.T.C. sponsor is the Forks Coffee Shop in downtown Forks."

Interorbital Systems, run by Rod and Randa Milliron out of an unmarked building at the Mojave airport, is also severely underfinanced. Rod is a rocket engineer who once worked for Grumman Aerospace, on Long Island. The Millirons dress in black, and they maintain a long-standing interest in electronic music. They spend part of their time in Hollywood, appearing as extras in movies and on television: Rod has a recurring nonspeaking role as a counter-terrorism agent on *24*. When they're in Mojave, they work on designs for space-faring vehicles. In a conservative town — a sign outside the local liquor store reads ICE AND AMMO — the Millirons are unsettling neighbors. "If aliens came down to Earth and were looking for company, I think they'd like to meet Rod and Randa," Dick Rutan told me.

"We like to say we're the most hated X Prize team," Randa Milliron said.

Interorbital plans to build a 115-foot rocket, which would be launched from a site that the Millirons have examined on Tonga, in the South Pacific. The rocket's capsule would return to Earth by parachute. The Millirons are impatient with the idea of putting wings on a spaceship: a wing serves no purpose in a place without air, and it is bound to add weight. In the opinion of the Millirons, Burt Rutan's winged design is undermined by the fact that it does not point the way to orbital flight: they say it could never be sufficiently modified and powered to take part in the next stage of the private space race (a claim that Rutan disputes). "Orbital is where the money is going to be made," Randa Milliron told me when we met. In fact, this summer Interorbital was no longer actively planning for an X Prize attempt — although a Rutan setback and the sudden appearance of a wealthy investor could bring the Millirons back in. Rather, they were concentrating on their interest in sending tourists into orbit and, in time, building a private space station. A full program would require an investment of $35 million. Randa said that when she and her husband meet other X Prize competitors they are struck by the fact that "they seem to be more into winning a prize than traveling in space. I haven't met too many competitors who are interested in setting up a colony on the moon."

The Millirons consider Rutan to be a man for whom a short journey matters more than any dream of arrival. They are also highly suspicious of Rutan's X Prize progress. "It could all be preprogrammed to be a success, whatever happened," they told me, suggesting that one test flight by Rutan's team might not have achieved the height claimed for it. Rutan says of some of his competitors, whom he never refers to by name, "They start with a little bit of money hoping they'll make progress, and then they find themselves doing stunts instead of research flight tests. They beg for more money instead of putting their nose to the grindstone."

To speak with apparent certainty, as the Millirons do, of vacationing in a space hotel is to show extreme optimism in one of two areas: either in the speed at which the private space industry will mature or in the chances that one will live a very, very long time. The Millirons do have some more modest technological plans: they have built and test-fired engines for rockets intended to launch small satellites, for example. But they also claim to follow a regimen of diet and exercise dictated by their desire to undertake future space travel (a regimen that, unconventionally, includes milkshakes), and they allow themselves to be drawn into conversations about space etiquette: the wisdom of keeping goats in orbit, say, or making Venus habitable by forcing a greenhouse effect upon it. At such times, they risk sounding like a fringe presidential candidate musing on possible curtain colors for the Lincoln Bedroom.

Rutan's company, Scaled Composites, leases buildings a hundred yards from Interorbital. A drab light-industrial corridor leads past a number of glossy paintings that portray all of Rutan's aircraft flying in formation over lush landscapes. Beyond the coffee station and the cut-out *Dilbert* cartoons, a door opens onto a 22-foot-high hangar where, on a recent visit, Rutan showed me *SpaceShipOne*, his X Prize competitor. It looked as playful as a Philippe Starck hair dryer and as flimsy as a fiberglass canoe. One would feel nervous driving it around the corner to the airport restaurant, for fear of running into a jackrabbit. About 30 feet long, it looked like the scale model for something much larger, sturdier, and more intergalactic. *SpaceShipOne* was recognizably a plane — a white tapered tube with wings and an undercarriage, parked parallel to the ground — but the wings were hinged, and the small cabin space, where the tube narrowed and then came to a point, had a dozen portholes at different heights in place of a windshield. The flight deck featured a single computer screen. The pilot's controls were a stick and two foot pedals. The interior had an unpolished, homebuilt feel, and smelled of carbon.

Rutan showed me a Ping-Pong ball decorated with hand-drawn happy faces that hung from above the pilot's head; when the string is no longer taut, the pilot is weightless. "You don't fly the space shuttle to orbit," he said. "The pilot sits there and the *computer* flies it. They don't trust the pilot to fly it. If space is going to be cheap, it has to be stick-and-rudder."

On the morning of June 21, 2004, a year behind schedule, and after fourteen test flights that had stayed mostly under 20 miles, Mike Melvill, sixty-three, an old friend of Rutan's, and "the best stick-and-rudder guy" he knows (a "steely-eyed, hairy-chested test pilot," in the words of Tonya Rutan), strapped himself into the pilot's seat in *SpaceShipOne* with a parachute on his back. Melvill had flown *SpaceShipOne* nearly a dozen times, but this would be the first flight to aim for space and the last test before a formal attempt at the X Prize. At 6:47 A.M., *White Knight,* another Rutan design, with long skinny wings, lifted *SpaceShipOne* from the runway. An hour later, the twinned planes were at 47,000 feet. White Knight's flight-test engineer pulled the handle that released *SpaceShipOne,* and Melvill glided for a few seconds alone. He then ignited the rocket engine — burning a mixture of liquid nitrous oxide and rubber — and within 10 seconds *SpaceShipOne* was moving upward faster than the speed of sound.

Rutan later acknowledged that he had considered the risks of producing a "smoking hole" on the desert floor. "I'd have lost a friend. You could say, 'I should pick a pilot who I'm less friendly with, a guy who's a stranger to me and just working for me, so if he gets killed . . .'" He smiled. "You could say, 'Let's have a lawyer fly it'" — a pause — "'or a liberal.' Mike's a conservative and a good friend, so that's the highest risk I can take." Dick Rutan, who would gladly train to pilot *SpaceShipOne,* said, "It worries me, what Burt would go through, if something terrible happened. I worry that he's too confident. It might totally destroy him."

Alan Bond, a leading British rocket engineer and the head of Reaction Engines, recently told me that it would take a shotgun to get him to ride on *SpaceShipOne.* He fears that the X Prize will discredit the entrepreneurial space race if contestants die on CNN, live. "When these people do kill themselves, which they surely will, in large numbers, then companies like mine, which look to governments to support them, we'll suffer," he said. Bond has calculated that an amateur space flight costing $10 million — the kind of money an X Prize competitor might hope to have — would carry a high risk of disaster. He made the calculation by determining the current cost of designing, building, and exhaustively testing a vehicle like the X-15: $1.3 billion. "So what can you do with only ten million? You can work out what the failure probability is, and it comes out at thirty to forty percent."

After firing the engine, Melvill intended to fly *SpaceShipOne* directly upward. But the plane was caught by a violent wind shear; he overcompensated and, after zigzagging, ended up several degrees away from vertical. "A lot of pitching," he told the test's mission control in Mojave, by radio. The deviation cost him height, but according to Rutan, it was not especially dangerous. After 76 seconds, a timer shut down the engine. Melvill had reached a height of 34 miles. *SpaceShipOne* had such momentum in the thinning air that it continued to move at 2,300 miles an hour, now in silence. Melvill arranged *SpaceShipOne* into its folded, "feathered" position, with the wings tilted upward, in readiness for reentry. The horizon was 900 miles away; the sky was black and empty. "Holy mackerel," Melvill said. Three minutes after leaving *White Knight*, as *SpaceShipOne* was reaching the apogee of its flight, there was a moment of alarm: when Melvill pushed the switch to set the trim of the stabilizers, which keep the spacecraft from pitching, only one moved. Within 2 seconds, he had resorted to a backup control, and the trim was set. Now he had a minute and a half before Earth's air would disturb his peace. He put his hand into his flight-suit pocket and — fumbling, because time was short — took out a dozen M&M's and opened his hand. On Rutan's video footage of the cockpit, the candies float around Melvill's head. It's so quiet you can hear them hitting the windows.

Melvill later compared the noise of reentry to "somebody talking to you very, very sharply." In another momentary drama, the fairing around the engine's nozzle, softened by the heat of the motor, buckled. (According to Rutan, this presented no danger to the flight.) Rutan later said, "You know that if you jump off your roof and fall a couple of stories you're going to break your leg, but imagine falling ten, fifteen, twenty miles." In the most unnerving moment of the flight, Melvill was accelerating so fast that his body was experiencing pressure at five times the level of everyday gravity. "It's like a very exciting roller coaster," Rutan said. "It's ferocious for a moment during reentry, just enough to be a lot of fun." At 60,000 feet, Melvill straightened out the wings, and *SpaceShipOne* became a glider, returning to the Mojave airport 80 minutes after taking off. Melvill climbed onto the roof of *SpaceShipOne* and straddled it like Major Kong in *Dr. Strangelove*. His wife, who had not wanted him to make the flight, was waiting on the tarmac. Rutan's data showed that Melvill had exceeded 100 kilometers by only 400 feet. "If Mike had eaten a big breakfast, he wouldn't be an astronaut," he said.

Rutan, whose wife said that he had barely slept in the two weeks before the test flight (he remembers no such difficulties), drove out from mission control with Paul Allen to meet Melvill on the runway, taking a printed sign from someone in the crowd. It read *SpaceShipOne*, GOVERNMENT ZERO and had been printed by Ernest Hancock, a libertarian talk-radio producer from Arizona. (The antigovernment rhetoric associated with *SpaceShipOne* disregards, among other things, the fact that the Mojave airport's taxiways were improved this year with a $3.9-million grant from the FAA.)

"I was ecstatic," Rutan told me. "We have a few little things to clean up, like the X Prize." Three weeks later, Rutan announced that he would make a formal claim on the prize, whose rules demand notice of sixty days. The first flight would take place on September 29 and the second sometime in the following two weeks.

Several weeks after Melvill's flight, John Carmack launched his Black Armadillo on a short, unmanned test flight. The craft was expected to rise to 200 feet before slowly returning; instead, it reached 600 feet, ran out of propellant, and crashed to the ground. The same weekend, Space Transport Corporation's Rubicon 1 rocket exploded over the Pacific. The head of one of the three dummies on board was washed ashore; the company later offered it for sale on eBay.

"It showed the world that the little guy can do it," Rutan told me when we had dinner recently. "Years of manned spaceflight and there was less activity this year than there was in 1961! That's ridiculous. Why is NASA running an airline? They should be doing basic research and making it available to American industry."

We were at the house in Mojave where he has lived since 1989. A few years earlier, during his third marriage, Rutan had been persuaded to move to Palmdale, which is half an hour south of Mojave and is more worldly. There he kept a poodle and shaved off the long, full sideburns that he regards as his trademark. When he met

Tonya — his fourth wife and "the best one he's ever had," in the affectionate view of a longtime neighbor — he regrew his sideburns, moved back to Mojave, and helped design a new house just outside town, situated on a spectacular spot in the desert, in the midst of Joshua trees.

He met me at the door of a white concrete hexagonal pyramid. At the dirt roadside, Rutan's street number was written on the tail end of a cargo plane; it pointed up at an angle, as if it had crashed into the sand. Inside, the house had the air of the hideaway of a young Medellín kingpin. It was a single open space reaching to a skylight at the pyramid's point. On one wall, where you might hope to see a view of the surrounding desert, a mural showed three stone pyramids by a lake and, in the foreground, four figures standing on a promenade: a human couple, the Egyptian god Horus, and an alien with large eyes. Aspects of the mural's design correspond to ideas that Rutan has about the origin of the Egyptian pyramids, some of which he believes may have been built by a civilization earlier than ancient Egypt. The sound of the wind did not carry through the walls.

We ate sitting on a sofa beside a red trapezoid pool table, among stuffed toy animals that included a four-foot bear, a gift from Burt to Tonya. Rutan spoke in long, stump-speech paragraphs. When he wanted to emphasize a point, he leaned forward with the unsettling smile of a department-store Santa asking a child if he has been good. "No one went to jail!" he said of NASA's troubles with the space shuttle. "You go down to Kmart with a ski mask on for thirty-five dollars, and they send you to jail. But can you imagine spending hundreds of billions of dollars *after* making a promise to Congress that this will save money? No one went to jail!"

As we spoke, a bird screeched at increasing volume in a side room. Eventually, Tonya fetched Winglet, an African gray parrot that Burt had reared from its birth, twenty-five years ago. His tone softened.

"Can you do a duck?" Rutan asked the parrot. The parrot quacked.

"Can you do a cat?"

The parrot barked like a dog.

"A cat!"

It meowed.

Rutan picked up the bird and put its head into his mouth, as if to bite it off, and made a roaring sound. The bird clucked. "She likes the echo," Rutan said.

Rutan's desert house has no windows on the ground floor. It is a dark, silent capsule, designed for energy efficiency. Windows would compromise the home's protection from heat. Earlier, Tonya had described how her husband lived primarily in the company of his own thoughts, and the house seemed to be a symbol of this temperament, as well as a product of it. Like *SpaceShipOne* — a solution to an engineering problem solved forty years ago; a design to inspire the awe of Earthlings while moving at high speed away from them; a smallish step for mankind, a giant leap for Burt Rutan — the house's design reflects a mind that seems to be happiest when it ignores the world beyond it. "I see enough of the ugly desert when I drive to work," Rutan said, laughing.

A few years ago, Rutan developed heart trouble, and he was fitted with a defibrillator. His brother refers to him as Robo-Burt. He is now prohibited from flying a plane except with another pilot on board, and, according to Tonya, there is doubt whether he would risk riding as a passenger on a future flight of *SpaceShipOne* — not for fear of his own safety but out of worry that his flailing death by heart attack would distract the pilot and endanger the flight.

Yet whenever Rutan talks of space tourism he still includes himself in the adventure. He figures that he has forty or fifty more years to live. During dinner, he said, "I really think I can go to the moon now. There's a whole bunch of people with new dreams. And I think they're valid now." He takes calls every day from people wanting to buy tickets to space from him; he explains several times a day that he has built a prototype and has not launched an airline. Yet he thinks that space tourism will soon be economically viable. "You could charge someone a hundred thousand dollars, maybe more. To get rid of the rich guys, service the rich guys. Theoretically, I could do it next year. But I don't call *that* space tourism. Space tourism is when you get more seats, and you can do it at thirty or fifty thousand dollars. And then the next generation would be ten or twelve thousand. Nobody's space airline will be as safe as an airplane. But it's going to be as safe as skydiving." He looked at the giant teddy bear, which he has half-seriously considered as a *SpaceShipOne* passenger on September 29, and added, "Well, certainly safer than climbing Everest."

OLIVER SACKS

In the River of Consciousness

FROM *The New York Review of Books*

1

"TIME," says Jorge Luis Borges, "is the substance I am made of.
Time is a river that carries me away, but I am the river . . ." Our
movements, our actions, are extended in time, as are our perceptions, our thoughts, the contents of consciousness. We live in time,
we organize time, we are time creatures through and through. But
is the time we live in, or live by, continuous — like Borges's river?
Or is it more comparable to a chain or a train, a succession of discrete moments, like beads on a string?

David Hume, in the eighteenth century, favored the idea of discrete moments, and for him the mind was "nothing but a bundle or
collection of different perceptions, which succeed each other with
an inconceivable rapidity, and are in a perpetual flux and movement."

For William James, writing his *Principles of Psychology* in 1890, the
"Humean view," as he called it, was both powerful and vexing. It
seemed counterintuitive, as a start. In his famous chapter on "the
stream of thought," James stressed that to its possessor, consciousness seems to be always continuous, "without breach, crack, or division," never "chopped up, into bits." The content of consciousness
might be changing continually, but we move smoothly from one
thought to another, one percept to another, without interruption
or breaks. For James, thought *flowed;* hence his introduction of the
term "stream of consciousness." But, he wondered, "is consciousness really discontinuous . . . and does it only seem continuous to itself by an illusion analogous to that of the zoetrope?"

Before 1830, there was no way (short of making an actual working model, or toy theater) of making representations or images that had movement. Nor would it have occurred to anyone that a sensation or illusion of movement *could* be conveyed by still pictures. How could pictures convey movement if they had none themselves? The very idea was paradoxical, a contradiction. But the zoetrope proved that individual images could be fused in the brain to give an illusion of continuous motion, an idea that was soon to give rise to the motion picture.

Zoetropes (and many other similar devices, with a variety of names) were extremely popular in James's time, and few middle-class Victorian households were without one. All of these instruments contained a drum or disk on which a series of drawings — of animals moving, ball games, acrobats in motion, plants growing — was painted or pasted. The drawings could be viewed one at a time through radial slits in the drum, but when the drum was set into motion, the separate drawings flicked by in rapid succession, and at a critical speed this suddenly gave way to the perception of a single, steady moving picture. When one slowed the drum again, the illusion vanished. Though zoetropes were usually seen as toys, providing a magical illusion of motion, they were originally designed (often by scientists or philosophers) with a sense that they could serve a very serious purpose: to illuminate the mechanisms both of vision and of animal motion.

Had James been writing a few years later, he might indeed have used the analogy of a motion picture. A movie, with its taut stream of thematically connected images, its visual narrative integrated by the viewpoint and values of its director, is not at all a bad metaphor for the stream of consciousness itself. And the technical and conceptual devices of cinema — zooming, fading, dissolving, omission, allusion, association, and juxtaposition of all sorts — rather closely mimic (and perhaps are designed to mimic) the streamings and veerings of consciousness.

It is an analogy that Henri Bergson used twenty years later, in his 1908 book *Creative Evolution,* where he devoted an entire section to "The Cinematographic Mechanism of Thought, and the Mechanistic Illusion":

> We take snapshots, as it were, of the passing reality, and . . . we have only to string these on a becoming, . . . situated at the back of the apparatus

of knowledge, in order to imitate what there is that is characteristic in this becoming itself . . . We hardly do anything else than set going a kind of cinematograph inside us . . . The *mechanism of our ordinary knowledge is of a cinematographical kind.*

Were James and Bergson intuiting a truth in comparing visual perception — and, indeed, the flow of consciousness itself — to such a mechanism? Are the brain mechanisms that give coherence to perception and consciousness somehow analogous to motion picture cameras and projectors? Does the eye/brain actually "take" perceptual stills and somehow fuse them to give a sense of continuity and motion? No clear answer was forthcoming during their lifetimes.

There is a rare but dramatic neurological disturbance that a number of my patients have experienced during attacks of migraine, when they may lose the sense of visual continuity and motion and see instead a flickering series of "stills." The stills may be clear-cut and sharp, and succeed one another without superimposition or overlap, but more commonly they are somewhat blurred, as with a too-long photographic exposure, and they persist for so long that each is still visible when the next "frame" is seen, and three or four frames, the earlier ones progressively fainter, are apt to be superimposed on each other. While the effect is somewhat like that of a film (albeit an improperly shot and presented one, in which each exposure has been too long to freeze motion completely and the rate of presentation too slow to achieve fusion), it also resembles some of E. J. Marey's "chronophotographs" of the 1880s, in which one sees a whole array of photographic moments or time frames superimposed on a single plate.

I heard several accounts of such visual effects while working in the late 1960s with a large number of migraine patients, and when I wrote about this in my 1970 book *Migraine,* I noted that the rate of flickering in these episodes seemed to be between six and twelve per second. There might also be, in cases of migraine delirium, a flickering of kaleidoscopic patterns or hallucinations. (The flickering might then accelerate to restore the appearance of normal motion or of a continuously modulated hallucination.) Finding no good accounts of the phenomenon in the medical literature —

perhaps not entirely surprising, for such attacks are brief, rare, and not readily predicted or provoked — I used the term "cinemato-graphic" vision for them, for patients always compared them to films run too slow.

This was a startling visual phenomenon, for which, in the 1960s, there was no good physiological explanation. But I could not help wondering then whether visual perception might in a very real way be analogous to cinematography, taking in the visual environment in brief, instantaneous, static frames, or "stills," and then, under normal conditions, fusing these to give visual awareness its usual movement and continuity — a "fusion" that seemingly was failing to occur in the very abnormal conditions of these migraine attacks.

Such visual effects may also occur in certain seizures, as well as in intoxications (especially with hallucinogens such as LSD). And there are other visual effects that may occur. Moving objects may leave a smear or wake in the direction they move; images may re-peat themselves; and afterimages may be greatly prolonged. I have experienced this myself, following the drinking of *sakau,* a halluci-nogen and intoxicant popular in Micronesia. I described some of these effects in a journal, and later in my book *The Island of the Colorblind:* "Ghost petals ray out from a flower on our table, like a halo around it; when it is moved . . . it leaves a slight train, a visual smear . . . in its wake. Watching a palm waving, I see a succession of stills, like a film run too slow, its continuity no longer maintained."

I heard strikingly similar accounts in the late 1960s from some of my postencephalitic patients, when they were "awakened," and es-pecially overexcited, by taking the drug L-dopa. Some patients described cinematic vision; some described extraordinary "stand-stills," sometimes hours long, in which not only visual flow was ar-rested, but the stream of movement, of action, of thought itself.

These standstills were especially severe with one patient, Hester Y. Once I was called to the ward because Mrs. Y. had started a bath, and there was now a flood in the bathroom. I found her standing completely motionless in the middle of the flood.

She jumped when I touched her, and said, "What happened?"

"You tell me," I answered.

She said that she had started to run a bath for herself, and there was an inch of water in the tub . . . and then I touched her, and she suddenly realized that the tub must have run over and caused a flood. But she had been stuck, transfixed, at that perceptual moment when there was just an inch of water in the bath.

Such standstills showed that consciousness could be brought to a halt, stopped dead, for substantial periods, while automatic, nonconscious function — maintenance of posture or breathing, for example — continued as before.

Another striking example of perceptual standstill could be demonstrated with a common visual illusion, that of the Necker cube. Normally, when we look at this ambiguous perspective drawing of a cube, it switches perspective every few seconds, first seeming to project, then to recede, and no effort of will suffices to prevent this switching back and forth. The drawing itself does not change, nor does the retinal image. The switching is a cortical process, a conflict in consciousness itself, as it vacillates between alternative perceptual interpretations. This switching is seen in all normal subjects and can be observed with functional brain imaging. But a postencephalitic patient, during a standstill state, may see the same unchanging perspective for minutes or hours at a time.

The normal flow of consciousness, it seemed, could not only be fragmented, broken into small, snapshotlike bits, but could be suspended intermittently, for hours at a time. I found this even more puzzling and uncanny than cinematic vision, for it has been accepted almost axiomatically since the time of William James that consciousness, in its very nature, is ever changing and ever flowing; but now my own clinical experience had to cast doubt on even this.

Thus I was primed to be further fascinated when, in 1983, Josef Zihl and his colleagues in Munich published a single, very fully described case of motion blindness: a woman who became permanently unable to perceive motion following a stroke. (The stroke had damaged the highly specific areas of the visual cortex that physiologists have shown in experimental animals to be crucial for motion perception.) In this patient, whom they call L. M., there were "freeze frames" lasting several seconds, during which Mrs. M. would see a prolonged, motionless image and be visually unaware of any movement around her, though her flow of thought and perception was otherwise normal. For example, Mrs. M. might begin a conversation with a friend standing in front of her, but not be able to see her friend's lips moving or facial expressions changing. And if the friend moved around behind her, Mrs. M. might continue to

"see" him in front of her, even though his voice now came from behind. She might see a car "frozen" a considerable distance from
her, but find, when she tried to cross the road, that it was now almost upon her; she would see a "glacier," a frozen arc of tea coming
from the spout of the teapot, but then realize that she had over-
filled the cup and that there was now a puddle of tea on the table.
Such a condition was utterly bewildering, and sometimes quite
dangerous.

There are clear differences between cinematic vision and the
sort of motion blindness described by Zihl, and perhaps between
these and the very long visual and sometimes global freezes experienced by some postencephalitic patients. These differences imply
that there must be a number of different mechanisms or systems
for the perception of visual motion and the continuity of visual
consciousness — and this accords with evidence obtained from
perceptual and psychological experiments. Some or all of these
mechanisms may fail to work as they should in certain intoxications, some attacks of migraine, and some forms of brain damage
— but can they also reveal themselves under normal conditions?

An obvious example springs to mind, which many of us have
seen and perhaps puzzled over when watching evenly rotating
objects — fans, wheels, propeller blades — or when walking past
fences or palings, when the normal continuity of motion seems to
be interrupted. Thus, occasionally, as I lie in bed looking up at my
ceiling fan, the blades seem suddenly to reverse direction for a few
seconds, and then to return equally suddenly to their original forward motion. Sometimes the fan seems to hover or stall, and sometimes to develop additional blades or dark bands broader than the
blades.

It is similar to what happens when, in a film, the wheels of stage-
coaches sometimes appear to be going slowly backward or scarcely
moving. This wagon-wheel illusion, as it is called, reflects a lack of
synchronization between the rate of filming and that of the rotating wheels. But I can have a real-life wagon-wheel illusion even
when I look at my fan with the morning sun flooding into my
room, bathing everything in a continuous, even light. Is there,
then, some flickering or lack of synchronization in my own perceptual mechanisms — analogous, again, to the action of a movie
camera?

Dale Purves and his colleagues at Duke University have explored wagon-wheel illusions in great detail, and they have confirmed that this type of illusion or misperception is universal among their subjects. Having excluded any other cause of discontinuity (intermittent lighting, eye movements, and so on) they conclude that the visual system processes information "in sequential episodes," at the rate of three to twenty such episodes per second. Normally, these sequential images are experienced as an unbroken perceptual flow. Indeed, Purves and his colleagues suggest, we may find movies convincing precisely because we ourselves break up time and reality much as a movie camera does, into discrete frames, which we then reassemble into an apparently continuous flow.

In Purves's view, it is precisely this decomposition of what we see into a succession of moments that enables the brain to detect and compute motion, for all it has to do is to note the differing positions of objects between successive "frames," and from these calculate the direction and speed of motion.

2

But this is not enough. We do not merely calculate movement as a robot might — we *perceive* it. We perceive motion, just as we perceive color or depth, as a unique qualitative experience that is vital to our visual awareness and consciousness. Something beyond our understanding occurs in the genesis of qualia, the transformation of an objective cerebral computation to a subjective experience. Philosophers argue endlessly over how these transformations occur, and whether we will ever be capable of understanding them. Neuroscientists, by and large, are content for the moment to accept that they do occur and to devote themselves to finding the underlying basis or "neural correlates" of consciousness, starting from such elemental forms of consciousness as the perception of motion.

James dreamed of zoetropes as a metaphor for the conscious brain, Bergson of cinematography — but these were, of necessity, no more than tantalizing analogies and images. It has been only in the last twenty or thirty years that neuroscience could even start to address such issues as the neural basis of consciousness.

Indeed, from having been an almost untouchable subject before the 1970s, the neuroscientific study of consciousness has now become a central concern, one that engages scientists all over the world. Every level of consciousness is now being explored, from the most elemental perceptual mechanisms (mechanisms common to many animals besides ourselves) to the higher reaches of memory, imagery, and self-reflective consciousness.

It is now possible to monitor simultaneously the activities of a hundred or more individual neurons in the brain, and to do this in unanesthetized animals given simple perceptual and mental tasks. We can examine the activity and interactions of large areas of the brain by means of imaging techniques like functional MRIs and PET scans, and such noninvasive techniques can be used with human subjects to see which areas of the brain are activated in complex mental activities.

In addition to physiological studies, there is the relatively new realm of computerized neural modeling, using populations or networks of virtual neurons, and seeing how these organize themselves in response to various stimuli and constraints.

All of these approaches, along with concepts not available to earlier generations, now combine to make the quest for the neural correlates of consciousness the most fundamental and exciting adventure in neuroscience today. A crucial innovation has been "population-thinking," thinking in terms that take account of the brain's huge population of neurons (100 billion or so), and the power of experience to differentially alter the strengths of connections between them, and to promote the formation of functional groups or constellations of neurons throughout the brain — groups whose interactions serve to categorize experience.

Instead of seeing the brain as rigid, fixed in mode, programmed like a computer, there is now a much more biological and powerful notion of "experiential selection," of experience literally shaping the connectivity and function of the brain (within genetic, anatomical, and physiological limits, of course).

Such a selection of neuronal groups (groups consisting of perhaps a thousand or so individual neurons), and its effect on shaping the brain over the lifetime of an individual, is seen as analogous to the role of natural selection in the evolution of species; hence Gerald M. Edelman, who was a pioneer in such thinking in the 1970s, speaks of "neural Darwinism." The French neuroscientist J. P. Changeux is more concerned with the connections of individual neurons, and speaks of "the Darwinism of synapses." Both Changeux and Edelman will soon publish highly readable, general accounts of their work.

William James himself always insisted that consciousness was not a "thing" but a "process." The neural basis of these processes, for Edelman, is one of dynamic interaction between neuronal groups in different areas of the cortex (and between the cortex and the thalamus, and other parts of the brain). He speaks here of "re-entrant" (that is, reciprocal) interactions, and sees consciousness as arising from the enormous number of such interactions between memory systems in the anterior parts of the brain and systems concerned with perceptual categorization in the posterior parts of the brain.

Other pioneers in the study of the neural basis of consciousness are Francis Crick (of the "double helix") and his younger colleague Christof Koch, who, from their first collaborative work in the 1980s, have focused more narrowly on elementary visual perception and processes. Koch gives a detailed but vivid and personal history of their work, and of the search for the neural basis of consciousness generally, in his new book, *The Quest for Consciousness*. Mechanisms of visual consciousness, Crick and Koch feel, are an ideal starting point, because they are the most amenable to investigation at present and can serve as a model for investigating and understanding higher and higher forms of consciousness.

In a synoptic paper called "A Framework for Consciousness," published in *Nature Neuroscience* in February 2003, Crick and Koch speculate on the neural correlates of motion perception, how visual continuity is perceived or constructed, and, by extension, the seeming continuity of consciousness itself. They propose that "conscious awareness [for vision] is a series of static snapshots, with motion 'painted' on them . . . [and] that perception occurs in discrete epochs."

I was startled when I first came across this passage a few months ago, because their formulation seemed to rest upon the same notion of consciousness that James and Bergson had intimated a century ago and that had been in my mind since I first heard accounts of cinematic vision from my migraine patients in the 1960s. Here, however, was something more, a possible substrate for consciousness based in neuronal activity.

But the "snapshots" that Crick and Koch postulate are not uniform, like cinematic ones. The duration of successive snapshots, they feel, is not likely to be constant; moreover, the time of a snapshot for shape, say, may not coincide with one for color. While this "snapshotting" mechanism for visual sensory inputs is probably a fairly simple and automatic one, a relatively low-order neural mechanism, each visual percept must include a great number of visual attributes, all of which are bound together on some preconscious level. How, then, are the various snapshots "assembled" to achieve apparent continuity, and how do they reach the level of consciousness?

While a particular motion, for example, may be represented by neurons firing at a particular rate in the motion centers of the visual cortex, this is only the beginning of an elaborate process. To reach consciousness, this neuronal firing, or some higher representation of it, must cross a certain threshold of intensity and be maintained above it — consciousness, for Crick and Koch, is a threshold phenomenon. To do that, this group of neurons must engage other parts of the brain (usually in the frontal lobes) and ally itself with millions of other neurons to form a "coalition." Such coalitions, they conceive, can form and dissolve in a fraction of a second, and involve reciprocal connections between the visual cortex and many other areas of the brain. These neural coalitions in different parts of the brain "talk" to one another in a continuous back-and-forth interaction. A single conscious visual percept may thus entail the parallel and mutually influencing activities of billions of nerve cells.

Finally, the activity of a coalition, or coalition of coalitions, if it is to reach consciousness, must not only cross a threshold of intensity, but must be held there for a certain time — roughly a hundred milliseconds. This is the duration of a "perceptual moment."

To explain the apparent continuity of visual consciousness, Crick and Koch suggest that the activity of the coalition shows "hysteresis," that is, a persistence outlasting the stimulus. This notion is very similar, in a way, to the "persistence of vision" theories advanced in the nineteenth century. In his *Physiological Optics* of 1860, Hermann Helmholtz wrote, "All that is necessary is that the repetition of the impression shall be fast enough for the after-effect of one impression not to have died down perceptibly before the next one comes." Helmholtz and his contemporaries supposed that this aftereffect occurred in the retina, but for Crick and Koch it occurs in the coalitions of neurons in the cortex. The sense of continuity, in other words, results from the continuous overlapping of successive perceptual moments. It may be that the forms of cinematographic vision I have described — with either sharply separated stills or blurred and overlapping ones — represent abnormalities of excitability in the coalitions, with either too much or too little hysteresis.

Vision, in ordinary circumstances, is seamless and gives no indication of the underlying processes on which it depends. It has to be decomposed, experimentally or in neurological disorders, to show the elements that compose it. Thus it is decomposed vision — the flickering, perseverative, time-blurred images experienced in certain intoxications or severe migraines — which above all lends credence to the notion that consciousness is composed of discrete moments.

Whatever the mechanism, the fusing of discrete visual frames or snapshots is a prerequisite for continuity, for a flowing, mobile consciousness. Such a dynamic consciousness probably first arose in reptiles a quarter of a billion years ago. It seems probable that no such stream of consciousness exists in an amphibian, like a frog, which shows no active attention, and no visual following of events. The frog does not have a visual world or visual consciousness as we know it, only a purely automatic ability to recognize an insectlike object if this enters its visual field and to dart out its tongue in response. It has been said that a frog's vision is, in effect, no more than a fly-catching mechanism.

If a dynamic, flowing consciousness allows, at the lowest level, a continuous, active scanning or looking, it allows, at a higher level, the interaction of perception and memory, of present and past. And such a "primary" consciousness, as Edelman puts it, is highly efficacious, highly adaptive, in the struggle for life.

From such a relatively simple primary consciousness, we leap to human consciousness, with the advent of language and self-consciousness and an explicit sense of the past and the future. And it is this that gives a thematic and personal continuity to the consciousness of every individual. As I write I am sitting at a café on Seventh Avenue, watching the world go by. My attention and focus dart to and fro — a girl in a red dress goes by, a man walking a funny dog, the sun (at last!) emerging from the clouds. These are all events that catch my attention for a moment as they happen. Why, out of a thousand possible perceptions, are these the ones I seize upon? Reflections, memories, associations lie behind them. For consciousness is always active and selective — charged with feelings and meanings uniquely our own, informing our choices and interfusing our perceptions. So it is not just Seventh Avenue that I see, but *my* Seventh Avenue, marked by my own selfhood and identity.

Christopher Isherwood starts his *Berlin Diary* with an extended photographic simile: "I am a camera with its shutter open, quite passive, recording, not thinking. Recording the man shaving at the window opposite and the woman in the kimono washing her hair. Some day, all this will have to be developed, carefully printed, fixed." But we deceive ourselves if we imagine that we can ever be passive, impartial observers. Every perception, every scene, is shaped by us, whether we intend it, know it, or not. We are the directors of the film we are making — but we are, equally, its subjects too: every frame, every moment, is us, is ours — our forms (as Proust says) are outlined in each one, even if we have no existence, no reality, other than this.

But how then do our frames, our momentary moments, hold together? How, if there is only transience, do we achieve continuity? Our passing thoughts, as James says (in an image which smacks of cowboy life in the 1880s) do not wander round like wild cattle. Each one is owned, our own, and bears the brand of this ownership, and each thought, in James's words, is born an owner of the thoughts that went before, and "dies owned, transmitting whatever it realized as its Self to its own later proprietor."

So it is not just perceptual moments, simple physiological moments — though these underlie everything else — but moments of an essentially personal kind, which seem to constitute our very being. Finally, then, we come around to Proust's image, itself slightly reminiscent of photography (and even of Hume), that we consist entirely of "a collection of moments," even though these flow into one another like Borges's river.

MICHAEL SPECTER

Miracle in a Bottle

FROM *The New Yorker*

ONE DAY LAST SEPTEMBER, as Britney Spears was about to
board a flight to Los Angeles from London, a rectangular blue bottle fell out of her purse. She quickly stuffed it back in, but not before the paparazzi recorded the event. Neither Spears nor her
spokesman was willing to comment on the contents of the bottle,
but the next morning London's *Daily Express* published a page of
pictures under the headline EXCLUSIVE: POP PRINCESS SPOT-
TED AT AIRPORT WITH POT OF SLIMMING TABLETS. Spears
was apparently carrying Zantrex-3, one of the most popular weight-
loss supplements currently sold in the United States. The pill,
which retails at about fifty dollars for a month's supply, contains a
huge dose of caffeine, some green tea, and three common South
American herbs that also act as stimulants. It hit the U.S. market
last March and has had a success that would be hard to overstate.
Millions of bottles have been sold, and during the Christmas season it was displayed in the windows of the nation's largest chain of
vitamin shops, GNC. (It is so highly sought after that many of the
stores keep it in locked cabinets.) Zantrex-3 is also sold at CVS, Rite
Aid, Wal-Mart, and other chains, and over the telephone and on
the Internet. If you type "Zantrex" into Google, more than a hundred thousand citations will appear. At any moment, there are
scores of people auctioning the stuff on eBay.

Perhaps the most interesting element of Zantrex-3's success
story, however, is that it is far from unique. There are hundreds of
similar products on the market today, and they are bought by millions of Americans. And though Zantrex's manufacturer makes
some heady claims ("the most advanced weight control compound

period"), so do the people who sell Stacker 2 and Anorex (whose publicity assures us that the "genetic link" to obesity means that repeated diet failure is "not your fault"), along with those who sell Carb Eliminator and Fat Eliminator. Almost all of these compounds suggest that they can help people lose weight and regain lost vigor, and often without diet, exercise, or any other effort.

The diet pill business may be the most visible segment of the vitamin, mineral, and herbal supplement industry, but it is by no means the largest. Thousands of different tablets, elixirs, potions, and pills are sold in the United States, and remarkably little is known about most of them. That doesn't deter consumers. Since 1994, when Congress passed a law that deregulated the supplement industry and opened it to a flood of new products, the use of largely unproved herbal remedies — from blueberry extract for impaired vision to saw palmetto for the treatment of enlarged prostates and echinacea to prevent colds — has increased as rapidly as the use of any commonly prescribed drug.

Since that legislation, the Dietary Supplement Health and Education Act, became law, companies have been able to say nearly anything they want about the potential health benefits of what they sell. As long as they don't blatantly lie or claim to have a cure for a specific disease, such as cancer, diabetes, or AIDS, they can assert — without providing evidence — that a product is designed to support a healthy heart (CardiAll, for example), protect cells from damage (Liverite), or improve the function of a compromised immune system (Resist). There are almost no standards that regulate how the pills are made, and they receive almost no scrutiny once they are, so consumers never truly know what they are getting. Companies are not required to prove that products are effective, or even safe, before they are put on the market.

Still, there is more to the growing reliance on supplements than the lapses of a single law: Americans long ago wearied of taking doctors' orders, and, increasingly, they are skeptical about the motives of big pharmaceutical companies. People want to feel in control of their own health. Supplements, with their "natural" connotations and cultivated image of self-reliance, let them do that. There is even a word to describe all the things — other than plain food — that people consume in the pursuit of health: nutraceutical.

Nutraceuticals are found everywhere today, in foods fortified with "extra" vitamins, in sports drinks, in "enriched" water, and now even in candy. Six out of ten adults in the United States take one or more supplements each day. Often these include multivitamins, which are frequently recommended by physicians, but a staggering number of amino acids, weight-loss cures, and herbal tonics are also swallowed every day, all in the belief that they will improve health, fend off disease, or make up for dietary and behavioral habits that have placed obesity and indolence among the leading health problems facing the United States. Last year Americans spent $19 billion on dietary supplements — nearly five times as much as they did just a decade ago. And they spent that money on everything from Vitamin C to garlic (the uses of which vary, with benefits that are never clear), from kava (which the FDA says may cause severe liver damage but which is still widely available in health food stores as a remedy for stress) to comfrey (an herb of dubious value commonly used to quell irritated stomachs), and even ephedra, which the federal government only recently decided to ban, despite reports over the last eight years implicating it in scores of deaths and hundreds of strokes, seizures, and other severe maladies.

"For many people, this whole thing is about much more than taking their vitamins," Loren D. Israelsen, the executive director of the Utah Natural Products Alliance, and a principal architect of the 1994 legislation, told me not long ago. "This is really a belief system, almost a religion. Americans believe they have the right to address their health problems in the way that seems most useful to them. Often that means supplements. When the public senses that the government is trying to limit its access to this kind of thing, it always reacts with remarkable anger — people are even willing to shoulder a rifle over it. They are ready to believe anything if it brings them a little hope." Frequently, such products come veiled in a cloak of science. Ads for Zantrex-3, for example, claim that its "superior power is validated by a direct comparison of published medical studies . . . scientific fact . . . irrefutable clinical data." The people who sell the pills on the telephone don't rely on science at all, however, when they tell callers that the capsules in those blue bottles could change their lives.

"When I train salespeople, I say to them, 'Do you know what people are calling you for? It isn't the pill. They are calling you for

hope. That is really what they want from you,'" Don Atkinson, who
is the vice president of sales for Basic Research, the privately held
conglomerate that distributes Zantrex-3, told me recently. I spoke
with him in his office in Salt Lake City, which regards itself as
the Silicon Valley of the dietary supplement industry. Atkinson, a
hearty and engaging man with a graying buzz cut and a firm hand-
shake, slowly wrote the word "hope" on a lined piece of paper soon
after I came in. "The customer has been overweight for years. And
they have tried everything. And they have been on Atkins and everything else and nothing has worked. And some of these people
are so incapacitated by their weight and their problems associated
with it that they would like to die. Just wish they could just die. And
they dial up and they are unhappy people. And they think, OK, if I
take this and it doesn't work it's further evidence that I am a failure. Our job is to give them hope. To say, 'You know what? You can
do this.'" Atkinson stopped for a moment and pumped his right fist
in the air. "I love my job," he said. "And do you know why? Because
when I get up in the morning I know that somebody's life is better
because we are here. Somebody today got some hope."

Atkinson told me that he was delighted by the Britney Spears
news, not so much because of the publicity windfall but for the
larger message it conveyed. "You know what is great about that? It's
the fact that she is using a weight-loss product and she looks terri-
fic. Just the fact that we are even talking about what Britney Spears
uses or doesn't use to keep her weight down tells the whole wide
world that it's OK to be a little overweight and it's OK to work on it.
And it's OK to use things to help you get there. That's what it all
says to me, and that is why we are here."

Herbs have been ingested regularly, in every conceivable combination, for thousands of years, and many are clearly beneficial. Vitamins and minerals are essential for human health: calcium supplements have prevented perhaps millions of cases of osteoporosis;
folic acid helps prevent neural tube defects; insufficient amounts
of vitamin B_{12} can lead to dementia. Simply eating citrus fruit, and
the vitamin C it contains, was enough to vanquish scurvy, which in
the mid-eighteenth century killed more British sailors than the
wars that Britain fought.

The way that nutrients work in foods has come to be properly understood only in the past hundred years. Because of the lack of specific knowledge, the preceding era had been one of open and unapologetic quackery. Throughout the eighteenth and nineteenth centuries, patent medicine men roamed the United States. For every illness imaginable, they promised wondrous products and magical cures.

The rapid growth of patent medicines was largely a result of two unrelated events, and they eerily foreshadowed the 1994 Dietary Supplement Act and the rise of the Internet as a commercial tool. In 1793 Congress passed patent legislation that permitted manufacturers to protect their formulas (without requiring that they even work). Around the same time, the number of newspapers published in the United States began to increase dramatically. By the beginning of the twentieth century, the patent medicine business accounted for more newspaper advertising than any other kind of product. Many manufacturers became rich; some became famous. Lydia E. Pinkham's Vegetable Compound, advertised as "A Positive Cure" for "all those Painful Complaints and Weaknesses so common to our best female population," made Pinkham's face as recognizable then as Martha Stewart's is today.

In 1914 officials of the American Medical Association decided to analyze Pinkham's compound. It turned out to be 20 percent pure alcohol and 80 percent common vegetable extracts. Most patent medicines had similar ingredients. Sometimes they were laced with cocaine, caffeine, opium, or even morphine. It's not surprising that they provided a few hours' worth of relief. There was no restriction on the vast armamentarium of remedies on the market until 1906, when the Pure Food and Drug Act was passed, mainly as a result of the revelations in Upton Sinclair's book *The Jungle*. The act permitted the Bureau of Chemistry, which preceded the Food and Drug Administration, to ensure that labels contained no false or misleading advertising. For a while, at least, snake-oil salesmen went the way of the Conestoga wagon.

Since then, the pendulum has swung between unregulated anarchy and restrictions that outrage many Americans. It has usually taken a disaster to persuade Congress to adopt strict regulations. Sulfanilamide, a drug prescribed to treat streptococcal infections, was used safely and effectively for years in tablet and powder form. Most children can't swallow pills, though, and in June 1937 researchers at one company found that the drug would dissolve in diethylene glycol; they tested the mixture for flavor, appearance, and fragrance — but not for toxicity — and then shipped it all over the country. They overlooked one important characteristic of the solution: diethylene glycol, normally used as an antifreeze, is a deadly poison. Within weeks, scores of children were dead. The victims experienced severe abdominal pain, nausea, vomiting, stupor, and convulsions. In a letter to President Franklin D. Roosevelt, one woman described the death of her child: "Even the memory of her is mixed with sorrow for we can see her little body tossing to and fro and hear that little voice screaming with pain and it seems as though it would drive me insane." The next year, after 137 deaths, Congress passed the Food, Drug and Cosmetic Act, which finally gave the FDA the authority it needed to regulate such products.

For many years afterward, there was little controversy about drugs or about dietary supplements, which mostly meant vitamins and minerals. "It didn't use to be so complicated," Annette Dickinson told me. She is the president of the Council for Responsible Nutrition, which is the most influential of the many groups that look after the interests of the supplement industry. Each year supplement manufacturers contribute millions of dollars to political candidates. The industry has been remarkably successful in arguing that because the First Amendment protects commercial speech, it can be used in defense of any claim that includes even a hint of truth. Dickinson is an aggressive supporter of supplements, yet she acknowledges that charlatans have proliferated wildly in the past decade, making her job, and that of most reputable manufacturers, much harder. In Dickinson's view, the industry would be better served if it returned its focus to the core nutrients — basic vitamin and mineral supplements. "In the beginning, you had your one-a-days, and there were minerals and herbal products, too. A drug was something intended to treat or cure a disease, and you needed to have proof that it could do those things; the line between foods and drugs was absolutely bright and clear. If you made a disease-related claim for something that was not approved, the FDA would come down on you like a ton of bricks. All through the forties and the fifties and the sixties, that was true."

By the middle of the 1970s, as the complex relationship between diet and health became more fully understood, the distinctions between foods, drugs, and supplements began to blur. First, with a major report issued in 1977 by the Senate Select Committee on Nutrition and Human Needs, and then with studies by the National Academy of Sciences and other research groups, the government started telling Americans to alter their diets if they wanted to have long and healthy lives. Advice about ways to reduce the risk of heart disease, diabetes, and many cancers and other chronic illnesses became routine: eat less salt and fat and add fiber and whole grains; eat more fruits and vegetables and watch the calories. Food companies were eager to promote many of their products as medically beneficial. It was illegal, however, to suggest that there was a relationship between the ingredients in a commercial food and the treatment or prevention of a disease. Then in 1984 the Kellogg Company launched a campaign, in conjunction with the National Cancer Institute, in which All-Bran cereal was used to illustrate how a low-fat, high-fiber diet might reduce the risk for certain types of cancer. These days it is almost impossible to pass by a supermarket shelf and not encounter such claims, but All-Bran was the first case in which a manufacturer issued a statement that was interpreted widely as "Eat this product because it will help prevent cancer." It led to the era of product labels and completely changed the way Americans think about not only foods but dietary supplements as well.

Since then, the English language has been stretched to its limits in the attempt to link products to health benefits. Even claims that are true may be irrelevant. Vitamin A, for example, is important for good vision — as supplements for sale in any health food store will tell you. Insufficient consumption of vitamin A causes hundreds of thousands of cases of blindness around the world each year, but not in the United States; here people don't have vision problems arising from a lack of vitamin A. Although statements advertising vitamin A for good vision may, like many others, be legally permissible, they are meaningless. "The laws allow manufacturers to make fine legalistic claims," Paul M. Coates, the director of the Office of Dietary Supplements at the National Institutes of Health, told me. "What we now have is an entire cottage industry of creative linguistics dedicated solely to selling these products." Instead of mentioning a disease (which in most cases would be illegal without FDA approval), companies make claims that a food can affect the structure or function of the body. Such claims can appear on any food, no matter how unhealthy it is. You cannot assert that a product "reduces" cholesterol, but you can certainly say that it "maintains healthy cholesterol levels." You cannot state that the herb echinacea cures anything, since it has never been shown to do that. But there is no prohibition on stating that it "has natural antibiotic actions" and is considered "an excellent herb for infections of all kinds." Ginkgo biloba has been recommended to Alzheimer's patients because it "supports memory function." Does it? Since research is not required before a supplement is released, there are not nearly enough data to know.

In a report published in the *Journal of the American Medical Association,* scientists compared the effects of echinacea with a placebo in treating colds. Echinacea is one of the most commonly used cold remedies in the United States. But the study, of more than four hundred children over a four-month period, showed that a placebo worked just as well and that children treated with echinacea were more likely to develop a rash than those who took nothing. Studies like that are rare, since they cost money that manufacturers are not required to spend. But they are at least as likely to disprove benefits as to confirm them. Ginseng has long been promoted as an energy booster, for example, yet the military, in studies of possible energy enhancements for troops, has found it worthless. Still, in my local health food store not long ago I saw more than a dozen supplements advertising the "fact" that ginseng improves energy.

"It was all done for crass commercial reasons," Marion Nestle told me. Nestle, the former chairman of the Department of Nutrition, Food Studies, and Public Health at New York University, is the author of *Food Politics,* which examines in detail the ways in which the food and supplement industries influence the nutritional policy of the United States and the health of its citizens. "In the name of health! The companies have masked it in an argument for freedom of speech. And look at some of the ways it all plays out. Obesity is an epidemic in our country. Is this helping? Not a bit." She went on, "I was staying in California this summer at the house of some friends. They had all sorts of health food products for kids and, to my surprise, among them were shark-shaped fruit snacks with vitamin C and gummy bears with echinacea. It's *candy* masquerading as something that will improve a child's health. It comes in one-ounce packages. Just the right size to throw in a lunch pail. It's brilliant marketing. Just brilliant."

One recent Harris poll found that most people believe that if a supplement is on the market it must have been approved by some government agency (not true); that manufacturers are prohibited from making claims for their products unless they have provided data to back those claims up (no such laws exist); and that companies are required to include warnings about potential risks and side effects (they aren't). "When something goes wrong, though, most people expect government health officials to find a solution," David A. Kessler told me. Kessler, who is the dean of the School of Medicine at the University of California at San Francisco, was the FDA commissioner when Congress passed the Dietary Supplement Act, which he adamantly opposed. "This is really the classic American ambivalence, and it has always been part of our nature," he said. "The view of most people is simple: I want access to everything and I want it now." The Federal Trade Commission — not the FDA — regulates supplement advertising. But the FTC is principally concerned with commerce, not science: it focuses on the content of the labels, not the content of the pills. Although since 1994 the agency has sued more than a hundred diet-pill companies, in 2002 it found that at least half of all weight-loss ads contained false or misleading statements. Despite its vigilance, the agency has an impossible job; for each success, ten new companies seem to appear.

When people get sick, Dr. Kessler pointed out, the refrain is always "'Where the hell is the FDA to protect me?' The supplement industry doesn't have to report adverse events, so the FDA doesn't have the data it needs to protect people. You cannot prove something is unsafe if you don't have the data. It's the ultimate Catch-22. It is also a colossal failure to protect the public health of this country."

Until a year ago, when Steve Belcher, a twenty-three-year-old pitcher for the Baltimore Orioles, died of heat stroke after taking an over-the-counter product that contained ephedra, it was by far the most popular supplement in the United States, bringing in a billion dollars a year and accounting for more than 10 percent of the supplement industry's annual sales. Ephedrine, the herb's active ingredient, boosts adrenaline, stresses the heart, raises blood pressure, and increases the rate of a person's metabolism. Derived from the Asian herb ma huang, it seems to help with short-term weight loss and with increasing physical stamina. When used in combination with caffeine, as it often is, ephedra is associated with an increased risk of heart attack, stroke tachycardias, palpitations, anxiety, psychosis, and death. Even though it was cited as a contributing factor in Belcher's death, only three states — New York, California, and Illinois — subsequently banned supplements containing ephedra. After numerous studies and nearly a year of review, the FDA announced, on December 30, that it would prohibit the sale of such supplements. Yet because ephedra is not a drug, it will be several months before the ruling works its way through the federal bureaucracy. (The agency has recognized for years that ephedra can be dangerous; its use in over-the-counter medicine has been regulated since 1983.)

Despite the risks, the appeal of diet pills is not hard to understand. Each year, obesity kills millions of Americans and costs billions of dollars. Data from the National Health and Nutrition Examination Survey show that almost 65 percent of the adult population is overweight. The prevalence of obesity among children is spreading, and if current trends continue more than 40 percent of Americans will be clinically obese within five years. The burden on the health system, not to mention the weakened quality of life that obesity causes, will be enormous.

I stumbled across an advertisement for Zantrex-3 while riding on the subway in New York one day. The name seemed coolly futuristic, and the ad was inviting, featuring an impossibly lithe and attractive couple dancing the tango. The copy beneath them promised — in "one amazing superpill" — both weight loss and incredible energy. "You are not obese," Zantrex-3 ads say. "You just need to lose a quick ten to fifteen pounds . . . and you want energy . . . plenty of energy."

Who could argue with that? I looked for the name of the company at the bottom of the ad: Zoller Laboratories. When I called the 800 number printed on the advertisement, the woman who answered told me that the company was based in Salt Lake City, but I couldn't find it listed in any of the databases that I normally use for research. Then I noticed, in an article about Britney Spears's "weight problems," that the chief scientist at Zoller was quoted by name. I dialed the 800 number again and asked to speak with him. He answered the phone, but was startled when I asked if I could fly out and talk with him. He promised to call me back. He never did, but eventually the public relations representative for a company called Basic Research invited me to visit Basic's headquarters, in Salt Lake City. He told me that Basic was the exclusive distributor of Zantrex-3. Zoller Laboratories does exist, but there are no offices and no labs. It's a company created by the marketing team at Basic because its name sounds scientific.

Basic Research is a privately held conglomerate based in a modern, 150,000-square-foot factory that was previously the U.S. assembly headquarters for Palm Pilot. Having tripled in size in the last year alone, the company is looking for more space nearby. The headquarters is a five-minute drive from the Salt Lake airport, and there are more than a dozen similarly squat industrial buildings scattered along the highway among the Days Inns and Guest Quarters. Most are home to companies with names like Utah Scientific, Cephalon, and Compeq. With the vigorous help of Senator Orrin G. Hatch, a principal sponsor of the Dietary Supplement Health and Education Act, the area has become a magnet for supplement companies. Hatch has been the industry's greatest champion and has consistently fought tighter regulations on products like ephedra. (So has his son, Scott, who has earned millions of dollars for firms that lobby on behalf of supplement companies.)

Arriving early for my meetings, I waited in the lobby, which was festooned with flyers and advertisements for a stupefying variety of herbal tonics and miracle cures. A banner bearing the company motto — "We help people feel great about themselves" — was stretched across an open bullpen area where dozens of salespeople worked the phones. Basic has a remarkably high "closure" rate; more than 60 percent of callers make a purchase. An advertisement for Zantrex-3 on the wall declared, "Dietary Supplement Industry Pinning Hopes on New 'Super Stimulant,' Non-Ephedra Diet Pill. Decline in Ephedra Diet Pill Sales Reversed by Zantrex™ 3's Sudden Popularity."

Basic puts out scores of products, which are marketed under the names of nearly a dozen companies — a practice that, according to Dennis Gay, the president and CEO, is intended to confuse competitors and "protect our brands in the Wild West atmosphere that exists today in the supplement industry." With Zantrex-3, Basic has seized cleverly on the fears about ephedra — marketing the pills as the "high-tech" substitute. But the company has many similar products; the cynically named Anorex, for example, is "the first weight-control compound designed to mitigate the profound effect that variations in the human genetic code have on the storage, use, and disposition of body fat," and Relacore is the "most significant weight-control advancement in more than a decade." ("Excess tummy flab is not your fault.") There is also Sövage Breast Augmentation Serum, a topically applied bust cream ("Yes They Are Real Breasts," one highly illustrative ad says), and Sövage Lip Plumper, to increase the fullness of one's lips. The company also markets tummy gels that promise "ripped abs" (NutraSport Cutting Gel), and a variety of tonics to help one think, relax, or sleep more soundly.

One product, Strivectin-sd, a cream that is sold for more than a hundred dollars a tube in Nordstrom, Lord & Taylor, and other stores, is made by a company called Klein-Becker USA, which calls itself "the industry leader in providing patented and exclusive weight-control and life-enhancement products that meet your individual needs." There is, however, neither a Klein nor a Becker, nor are there any specific employees who work there. Like Zoller Labs, it's a company created by Basic because the name sounds impressive and pharmaceutical. Strivectin was originally intended for use by women to reduce stretch marks (and in many stores it is still marketed that way), but people soon began to rub it on their faces as well; Strivectin now asks, without any data to justify the comparison or the question, "Even better than Botox?"

Daniel B. Mowrey, the man responsible for creating most of these products, is a gentle-looking figure with blue eyes and gray hair that is thinning at the top. The day I met him, he was dressed in chinos, a denim work shirt, sneakers, and a loud paisley tie. "I used to be a hippie," he said, shrugging, when he saw me staring at the tie. He told me he bought it long ago in Haight Ashbury. Mowrey is the director of scientific affairs at Basic and one of the three owners. Everyone calls him Dr. Dan. He received a Ph.D. in psychology from Brigham Young University, and although he never studied botany formally, he has written widely on the medicinal uses of herbs. He laid out his philosophy quite clearly in his book

The Scientific Validation of Herbal Medicine, published in 1986: "The scientific method is a powerful tool, but it has its limits . . . Medical science in America is a unique combination of economic and political factors, which fuse together almost religiously to promote synthesized, highly active chemicals." Mowrey told me he believes that there is almost always a "natural" alternative to synthetic drugs: "One that is cheaper, safer, and, often, more effective."

Mowrey came up with the components of Zantrex-3 the way he comes up with the elements of most of the company's products: by surfing the Internet. "I never understand why my competitors don't spend more time just looking at the information on the Web," he told me. "It's all out there," he said, showing me how he uses public databases — such as those kept by the National Institutes of Health — to see what's new in fields like weight control, memory, and aging.

Basic makes two major claims for Zantrex-3: that it will provide an immediate and sustained burst of energy and that it will help people lose weight rapidly. For the first claim, Mowrey relied on a study by the U.S. military that examined the effects of caffeine on navy SEALs who had been deprived of sleep and exposed to the extreme stresses of a training week. The study concluded that it is more effective in combating fatigue than a placebo. One dose of Zantrex-3 is like drinking four cups of strong coffee. Whatever the merits of that study, almost anyone who takes the pills is sure to feel the jolt — and many people buy them for that reason alone. (The clerk at my local GNC warned me to take the pill with food, or I might get "too high.")

The company attaches a tiny brochure to the neck of each bottle which says that Zantrex-3 caused "546 percent more weight loss" than America's No. 1 ephedrine-based diet pill — "without diet and exercise." It goes on, "Published clinical studies don't lie." I asked Mowrey to show me the data he used to arrive at that figure. He acknowledged that the figure was based not on a direct comparison of the two diet products but on extrapolations of results from unrelated studies.

One of the studies, which Mowrey describes as a "groundbreaking" paper, published in 2001 by two Danish researchers in the British *Journal of Human Nutrition and Dietetics,* evaluated the effectiveness of a mixture of three South American herbs — now used in

Zantrex-3 — in aiding weight loss by making people feel too full to eat. The study followed forty-six subjects for forty-five days. Half were given a placebo and the others received the herbal mix. At the end of the study, the herbal group had lost eleven pounds, on average, whereas the other group had lost less than one. Seven participants were followed for another year, during which they neither gained nor lost more weight.

The subject of obesity, after being largely ignored by the medical establishment, has finally gained currency in the United States, and many major medical schools and scientific institutes now are pouring research money into the field. Yet despite thousands of weight-loss studies and an increasingly focused search for solutions, there is no evidence that any prescription, over-the-counter product, or supplement has ever kept a person's weight down for much more than a few months. At best, such drugs or supplements are short-term answers to lifelong problems; at worst, they intensify the disorders they attempt to cure. I asked Mowrey if it was fair to assert, as he has, that "whether weight management or energy management is your goal, Zantrex-3 represents the very best options available anywhere."

"What options are better?" he asked. "We have to look at the study. We are not free to go beyond it, but it's not fair to ignore it, either."

Losing eleven pounds in forty-five days certainly sounds promising, but the results of a single six-week study involving fewer than fifty people would almost never provide enough meaningful data to prove the value of a drug or supplement of any kind. It usually takes years and involves hundreds, if not thousands, of subjects before a study of a new drug can yield clear evidence that it is effective. The main herbs in Zantrex-3 — guarana, yerba maté, and damiana, coupled with caffeine, zanthene, and green tea, among other ingredients — are stimulants, laxatives, and diuretics. You do not need a degree in nutritional sciences to realize that if you take a combination of stimulants, laxatives, and diuretics for six weeks you are going to lose weight, or to know that, over the long run, such a diet plan is certain to fail. "The idea that a pill, a mixture of herbs, or anything else will allow people to lose weight and keep it off without any other effort is completely ridiculous," Kelly Brownell told me when I called to ask his opinion of Zantrex-3. Dr.

Brownell is the chairman of the psychology department at Yale, and he is also the director of the Yale Center for Eating and Weight Disorders. "You look at a study that, in the end, followed seven people for a year and you can conclude nothing from that."

Mowrey argues that Americans ought to have the chance to make decisions about the value of supplements for themselves. "There are a lot of pharmaceuticals derived from plants," he said. "Lots of times, the safety issues are not important. And you have to remember what you have to do if you develop a drug today. Say you do a small study of maybe a hundred and fifty people and you find that as a result of the study eighty-five of the women who take this who would otherwise get breast cancer don't. The FDA demands that the company spend several billion dollars and fifteen years of research answering every little question that comes along. Every nitpicky little question. Now, how many people have you *killed* before you introduce this drug to the market?

"Drug companies don't offer money-back guarantees," he continued, emphasizing one of Basic's main policies. "We do. And if it isn't going to work, if it's not effective, then we have the ability to give money back. There is satisfaction guaranteed here. Can you imagine a drug company doing that? We are in the business of wellness, not of curing sick people. A lot of dietary supplements are designed to prevent problems from ever happening. There is no drug that is going to prevent illness. Drugs treat illness. They are going to be very, very invasive. Whereas dietary supplements are not invasive. You can combine vitamins with minerals and plants together in a thousand ways without anything happening that is bad."

The notion that herbal combinations are "natural" and therefore can't cause harm serves as a first principle for many people who take supplements as a solution to their medical problems. Even the most seemingly benign substances, however, can turn out to have significant and wholly unexpected effects. Perhaps the best example is grapefruit juice, which can disrupt the work of a series of enzymes that are found in the small intestine and which serve to break down drugs before they are absorbed into the bloodstream. Taking medicine with grapefruit juice permits it to enter the bloodstream in dangerously high concentrations, which keeps it from doing its job and can intensify many side effects. Many common pharmaceuticals — including antidepressants, antihistamines, and cholesterol medications — are not metabolized properly if they are taken with grapefruit juice.

There are numerous examples of herbs, drugs, and supplements that cause reactions or that when taken together are harmful. Beta-carotene, found in carrots, has always been considered purely beneficial, yet recent research has shown that, for men with certain types of cancer, those who took beta-carotene supplements had a significantly worse prognosis than those who did not. There is a scientific maxim that the dose makes the poison — that any substance, no matter how useful, can cause trouble if you take too much of it. Most physicians don't even know what supplements their patients are taking, let alone how much, so trying to warn people about possible interactions among them is impossible. "The remedy for all this is to stop dangerously pretending that pharmacologically active substances called dietary supplements should be treated completely differently from pharmacologically active substances called drugs," Sidney Wolfe, the director of Public Citizen's Health Research Group, told me. "You cannot determine if they are safe or effective without doing the studies. And with supplements the studies are almost never done."

Some herbs do work, of course, yet the absence of effective manufacturing standards in the United States means that even then consumers can't rely on commercial formulas. Black cohosh has been used for centuries to treat a variety of common ailments, including, most recently, menstrual and menopausal problems. In Europe it is considered a drug — and many studies have shown that it can have value. Women often take some form of the root instead of using hormone replacement therapy. Still, in the United States the herbal product that you buy tomorrow may be different biologically from the same product purchased next month. I have a friend who, at the onset of menopause, began to use a supplement that is composed principally of black cohosh. For several months her symptoms disappeared. One day, however, she bought a new bottle, and within a week her symptoms had returned so severely that she called her doctor from a car on the West Side Highway. "I demanded a prescription for hormone replacement therapy," she told me, even though she considers it dangerous. Her doctor guessed correctly that she had just bought a new bottle of the supplement and advised her to switch to a different product containing black cohosh — Remi-Femin — which is made by Shaper & Brunner in Germany, where it is regulated as a drug.

With herbal products made in the United States, however, there is simply no way to know what you are getting in each bottle. In 2002 researchers at the New York Botanical Garden published a study in the journal *Economic Botany* in which they reported on using DNA-fingerprinting techniques to identify several species of black cohosh. They found great variation in the herbal mixtures that were turned into products for the marketplace. It's hard to make a botanical product exactly the same way every time. Without rules, there is almost no incentive to try.

Since the standards for making diet pills are set largely by the people who sell them, I decided it would be useful to see how Zantrex-3 was made. "You go look at the factory," Dennis Gay told me. "If you think we are a sleazy operation, remember: we could do it for half the price." The next day I drove to Cornerstone Nutritional Labs in Farmington, about twenty minutes north of Salt Lake City on I-15. Cornerstone, an independent company that produces most of the Zantrex-3 sold in the United States, pumped out 1.7 billion capsules, tablets, and pills last year, nearly 200,000 an hour for every hour of every day. I was greeted warmly by Brent Davis, Cornerstone's director of nutraceutical sales, who offered to show me around the plant. We slipped on disposable booties, gowns, and hairnets, and removed the metal from our pockets, so as not to contaminate the materials. Our first stop was the receiving area at the plant's loading dock, where dozens of fifty-kilogram drums of raw herbs — green tea, yerba maté, and pangea among them — were lined up and stacked nearly to the ceiling, forty feet high. As soon as the herbs arrive at the factory, they are sampled for color, consistency, density, and purity. The raw materials are then taken to a weighing room, where they are collected by men in moon suits and sampled again. Most of the machines sit in clean rooms adjacent to the factory floor, cordoned off by walls of double-paned glass.

After the herbs are collected, they are mixed in a blender. This is not as easy as it might appear; natural organic compounds are far harder to combine than synthetically made drugs. A product like vitamin E comes in tiny balls, and most herbs come in flakes. The same herb can vary in consistency, in provenance, and even, at times, in species. Some need water; others are ruined by the slightest exposure to moisture. Some supplements require a minute amount of an ingredient — less than 150 micrograms, for instance — to be mixed evenly into more than 100 cubic feet of powder. And each supplement must be made in such a way that every capsule in every bottle is identical in quality and strength. "It's a hell of a job to do," said Michael Meade, the director of operations for Basic Research, who was also on the tour. "There's a recipe, and once it's worked out it's fine. But it takes time to get it right, and many companies fail. The idea that you just throw it all into the soup and wait is ridiculous." He said that Cornerstone was unusually rigorous in its testing, and that Basic was pleased by the consistency of the results.

Like other manufacturers in the secretive and, for the most part, privately owned supplement industry, Cornerstone declines to talk about its revenues or even name its clients. (Basic Research, too, reveals almost nothing about its earnings, expenditures, plans, or goals.) But Zantrex-3 is obviously a big part of Cornerstone's current business. The factory's largest blender, which was given over completely to the production of Zantrex-3 when I was there, can turn 5,000 kilograms of raw powder into the equivalent of 15 million pills a day. As soon as the newly homogenized herbal material leaves the blender, it is pressed by another machine into blue capsules, which are dumped into giant drums — 35,000 capsules in each drum. They are then collected in a hopper and fed into bottles.

Cornerstone's computer system monitors every gram as it passes down the line, and supervisors keep the floors spotless. Once the bottles are filled, they are capped and a tamperproof seal is melted on. Labels are applied by the same machine. Finally, one of the brochures advertising Zantrex-3's "amazing power" is fastened by hand to the neck of every bottle on the assembly line. From there the bottles are packed into boxes that are loaded into cartons, which are shrink-wrapped and ready to ship. If Basic Research were willing to cut a (totally legal) corner or two, there is no doubt that it could produce those pills for far less money.

*

When I left Cornerstone, I drove to a nearby Wal-Mart, which, along with such stores as Circuit City and Bed Bath & Beyond, anchors a mall in a suburb of Salt Lake City called Murray. Wal-Mart is the biggest of what Gay refers to as the "big boxes," the giant chains that can ensure a product's success simply by stocking it. By summer all 4,000 Wal-Mart stores will carry Zantrex-3; many of them will feature displays with the tango-dancing couple that I had noticed on the subway in New York.

The Murray Wal-Mart has an extensive section devoted to supplements of all kinds: bottles, packets, and cartons promising the usual array of unproved benefits and promoting the health of the eyes, the skeletal system, the urinary tract, the brain. A tiny asterisk appeared on every product — including Zantrex-3 — that suggested a connection between its contents and better health: "This statement has not been evaluated by the Food and Drug Administration. This product is not intended to diagnose, treat, cure or prevent any disease." If a product whose label promotes it as contributing to "wellness" is not intended to cure, treat, diagnose, or even prevent any health problem, what, one has to wonder, is it supposed to do? But there they all are, dozens of brands: Stacker 3, with chitosan, and Starch Away, which "blocks calories from bread, pasta, pastries and other foods" ("Dieting has never been easier," the bottle says). Zantrex was for sale, along with ZN-3, a new product that advertises itself on the label as being like Zantrex. It has the same ingredients and costs half as much, but since only the *names* of ingredients are listed, not the amounts, what comes in each capsule is anybody's guess.

As I walked out of the store, I heard an announcement on the loudspeaker: "Welcome, shoppers. Right now at the McDonald's inside this Wal-Mart you can get two cheeseburgers for only one dollar. This offer is for a limited time only. Also at this McDonald's we accept MasterCard and Visa."

Dennis Gay is a fifty-seven-year-old, pear-shaped man who has been waging an unsuccessful war on his weight for years. The first time I met with him, he was dressed in shades of green: olive pants, pale pine-colored shirt, and loden tie. He had a short, neatly trimmed graying beard. At that meeting, Gay was accompanied by a man who was described as a "consultant" but who obviously played a significant role in the company, because he did most of the talking. (Nearly every time I asked Gay a question, he deferred to his colleague.) The man spoke fast and had a New York accent. He didn't give me his name, although I asked him for it three times.

I wondered whether Basic Research believed that new laws requiring more regulation and stricter standards would be bad for the industry. "They are simply not needed," the man who wouldn't identify himself blurted out. "The FDA has the authority to act today if it wants to remove something from the market. But it always prefers to use the media." He went on to say that although he assumed that ephedra would be banned, there were really no significant problems with the herb, which Basic Research continues to sell, and which its salespeople told me on the telephone was the "best way by far to lose that weight fast." He said that Zantrex-3 increased "metabolism without increasing the heart rate or blood pressure" — which, as it happens, is almost never the case.

After the interview, I learned that the man's name was Mitchell K. Friedlander, and that he had made false claims about diet pills, and in the 1980s was prohibited by the U.S. Postal Service from selling them through the mail.

When I spoke to Gay a second time, Friedlander was not there. I said it was hard to ignore the fact that his most trusted adviser — Gay described him to me as a "marketing genius" — had been found to have made false claims about diet pills. Gay told me that he had hired a private investigator to "check Mitch out" and that he was comfortable with the report. "Mitch is valuable," Gay said. "He doesn't desire to become a part of the company, and I don't think we want him to be." I asked if working with a man with Friedlander's past bothered him, since he was trying to establish Basic Research as one of the more reputable companies in an extremely irregular business.

"Remember the old days when the banks hired safecrackers to protect them?" Gay said. "Is this that different? What are the standards in this industry? Tell me what they are, because I really wish I knew, so that I could abide by them. We try to do better, but there are no clear rules. None." He went on, "We would welcome standards as long as they don't take the choice away from the public. I wouldn't welcome the standards that exist on the drug side. Because then you have no choice. And the American consumer is not stupid. He deserves to make his own mind up about what he does."

This is Gay's bottom line, and that of the industry as well. "I have to get into my lecture," he said, and walked over to a whiteboard in the office. "Let's say I've got ninety-nine people that have a fatal form of cancer. The way the FDA regulates drugs now, a study would typically look like this." He drew three big circles and wrote the number 33 in each of them. "A third will get nothing, and they are going to die. Then another thirty-three we are going to give a placebo. The last third get the active ingredient they are testing as a new drug.

"Let's assume the first thirty-three die. What about placebo? Well, most studies show that placebo survival rate is thirty-six percent. That suggests that eleven of the people on placebo will survive at least long enough to be significant. And now let's say that of the third who receive the active ingredient eleven survive. Based on today's regulations, that drug would not be approved. It's no better than placebo. And it would be tossed in the trash. But this is what I want to ask: what about these twenty-two people here?" Gay drew a big line under the eleven placebo and eleven drug recipients who, in his reckoning, would have lived. "What the government of the United States says is that those twenty-two people don't have the benefit of a placebo or of the active ingredient. So you have zero people surviving out of ninety-nine when you could have had twenty-two surviving." Gay looked mournful. "All I am saying is that I want to have the right to appeal to those twenty-two people. I want to give them a chance to live."

There is a demonstrable placebo effect in most clinical studies, although the idea that placebo could save even eleven patients with fatal cancer is ludicrous. But Gay and Basic make their money by selling the dream of wellness, not the reality. If their products could really swell breasts, banish wrinkles, and erase fat, Basic would probably become the most successful company in American history. After all, is Zantrex-3 any different from Lydia Pinkham's miraculous concoctions?

This year Congress will consider a bill that would modify the 1994 law so that thousands of unregulated botanical substances would be treated more like drugs than like foods. Supplement manufacturers — and their customers — are preparing to fight any such change with every resource they can muster. The bill has been advertised as an assault on the First Amendment. The alarm has sounded across the Internet, and congressional offices have been besieged with protests. Walk into any health food store and you'll see leaflets warning that the government is about to deny you the right to choose your own fate.

Gay went back to his chair and sat silently for a moment. "We put disclaimers in our ads, and we give people the results of the studies and a money-back guarantee," he said. "What more could you want? Don't prevent people from using their own judgment. Let them try it. If it doesn't work, they can return it. That's what's fair. That's what's American."

CLIFF STOLL

The Curious History
of the First Pocket Calculator

FROM *Scientific American*

JOHANNES KEPLER, Isaac Newton, and Lord Kelvin all complained about the time they had to waste doing simple arithmetic. Foolscap covered with numbers obscured answers; elegant equations led to numerical drudgery. Oh, for a pocket calculator that could add, subtract, multiply, and divide! One with digital readouts and memory. A simple, finger-friendly interface.

But none were available until 1947. Then, for a quarter of a century, the finest pocket calculators came from Liechtenstein. In this diminutive land of Alpine scenery and tax shelters, Curt Herzstark built the most ingenious calculating machine ever to grace an engineer's hand: the Curta calculator.

Advertisements in the back pages of *Scientific American* in the 1960s promised an arithmetic panacea: "The Curta is a precision calculating machine for all arithmetical operations. Curta adds, subtracts, multiplies, divides, square and cube roots . . . and every other computation arising in science and commerce . . . Available on a trial basis. Price $125."

With its uncanny resemblance to a pepper grinder, this device — still owned by some lucky people — does everything that your ten-dollar pocket calculator can do. Except that it's entirely mechanical — no battery, no keypad, no liquid-crystal display. You turn a crank to add numbers.

A wind-up adding machine? You bet. Today I'm holding Curta in my left hand, grinding out answers with my right. To add, I enter numbers with little sliders, spin the crank, and the result appears in small windows circling the top. I'm literally crunching numbers.

And yes, it multiplies and divides — although I have to spin that crank ten or twenty times to find the product of two big numbers. There's no on-off switch, but a handy finger ring clears the memory. As for square and cube roots, well, you carry along special tables and remember a few shortcut algorithms.

This is no slide rule that approximates an answer to three or four places. Through the windows on the top, eleven numbers click into place. Hey — your electronic calculator probably can't deliver eleven digits of precision.

Okay, it does arithmetic. So why has the Curta been called "a treasure of our civilization" and "a marvel of technology"? Why do collectors cherish these devices when any cheap calculator works much faster?

Because along with its impressive arithmetical abilities comes the sensation of mechanical elegance and certainty: you set numbers by sliding dials that slip into place with little curtsies. The crank turns with the smoothness of a fine pocket watch. Digits click into position with neither slop nor drag. Each number is engraved in magnesium, and steel gears handle the computation. The Curta purrs as you calculate.

Then, too, this machine was designed to make calculating easy. To avoid errors, separate displays show the entered number, how many times you turn the crank, and the result. Detents — small catches — tell your fingers when each digit is entered and when the answer is ready. It's easy to undo an error, but a ratchet won't let you clash the gears by going backward. And you won't erase your answer by mistake, because the clearing ring can't be accidentally activated.

The Curta calculator combines the precision of a Swiss watch, the craftsmanship of an old Nikon F camera, and the elegance of a tango — all in a compact cylinder. In 1950 the Curta's portability startled engineers: a calculator that you could carry! All the more astonishing, then, that this device arose from the nadir of civilization, the Buchenwald concentration camp.

Just as today's technicians crave featherweight laptop computers, engineers and accountants long yearned for a portable mathematics machine. Thomas de Colmar built an adding machine the size of a piano for the 1855 Paris Exhibition. Fifty years later the Millionaire Calculator could not only add and subtract, it could directly multiply and divide. Yet it weighed more than sixty pounds. For a real pocket calculator, civilization would wait for Curt Herzstark.

Born in 1902, Herzstark grew up around calculators. His father sold Remington and Burroughs office machines in Vienna. Within a few years his family built a factory to make calculators. It thrived, and young Curt found himself demonstrating adding machines across Austria.

During World War I, his family's factory made war material. Afterward, with the factory's machines worn out or destroyed, his father took to selling used calculators until the factory could be rebuilt. At the same time, new competition appeared, including Fritz Walther, who had made automatic pistols but now found himself stymied by postwar disarmament. Seeing opportunity in office equipment, he converted his gun factory to one that made electric adding machines.

In the 1930s the calculator business multiplied. "But something was missing in the world market," Herzstark later recalled. "Wherever I went, competitors came with wonderful, expensive, big machines. I'd talk with a building foreman, an architect, or a customs officer who'd tell me, 'I need a machine that fits in my pocket and can calculate. I can't travel ten kilometers to the office just to add a row of numbers.'"

Manufacturers such as Monroe, Friden, and Marchant tried shrinking big desk models, like a watchmaker miniaturizing a clock into a wristwatch, but without much success. Visit an antiques store, and you may bump into a miniature calculator of yesteryear. The Marchant "lightweight" adding machine weighed thirty-four pounds, with nine columns of buttons and a carriage sporting eighteen mechanical readouts. Two big cranks sprout from the side, reminiscent of a Model T Ford. Accountants lugged them around in suitcases. That's what portable meant in 1935.

Having seen failed attempts to shrink desktop adding machines, Herzstark, then in his thirties, started anew. "I looked at everything backward. Let's pretend that I have already invented everything. What does this machine have to look like so that someone can use it?

"This machine can't be a cube or a ruler; it has to be a cylinder so that it can be held in one hand. And holding it in one hand, you would adjust it with the other hand, working the sides, top, and bottom. The answer could appear on the top."

Like a good software engineer, Herzstark began with the user interface, rather than letting the mechanism control the design. Instead of using a typewriter-style keyboard, he decided to wrap sliders around a cylinder so you would enter numbers by sliding a thumb or finger. This approach would create an area for the results register around the top of the cylinder, as well as a convenient site for a crank to power the calculator.

Other mechanical calculators used a separate mechanism to calculate every digit in the answer. For instance, the Friden calculator had ten columns of keys to enter a number, and it had ten separate sets of calculating gears — expensive and heavy. Herzstark realized that he needed only one calculating mechanism if it could be consecutively used by each input digit. His calculator might have eight sliders to enter digits, but the teeth (or steps) on a single central drum would handle the arithmetic. The drum would allow him to trim the size and weight of his machine.

By 1937 Herzstark understood how to perform arithmetic using a single rotating step drum. Everything would work for addition and multiplication. But his design stumbled over subtraction and division. He couldn't subtract by turning the crank backward, because adding two digits may create a carry condition after the operation, but subtraction requires a borrow beforehand. A single arithmetic step drum couldn't properly look ahead to see what might be coming.

"Traveling in a train through the Black Forest, I sat alone in a compartment. Looking out the window, I thought, 'Good grief! One can get the result of a subtraction by simply adding the complement of a number.'"

To find the nines complement of a number, just subtract each digit from nine. By adding a number to the complement of another number, you can simulate subtraction. For example, to calculate 788,139 minus 4,890, first find the nines complement of 004,890: 995,109. Now add 788,139 and 995,109 to get 1,783,248. Remove the highest-order digit to arrive at 783,248. Finally, add one to find the answer: 783,249. Sweet — the same technique is used in computers today.

Herzstark's calculator would retain the single rotating step drum but would have two sets of teeth: one set dedicated to addition, the other to subtraction. Lifting the crank 3 millimeters would engage the subtraction teeth to perform nines-complement addition. Subtraction would be as easy as addition.

Multiplication and division would be handled by repeated addition and subtraction. And since the results register could be rotated in relation to the input sliders, several shortcuts would speed these operations. For instance, to multiply by 31,415, you don't spin the crank 30,000-plus times; the movable carriage cuts this to fourteen turns: five turns for the 5, once for the 10, four times for the 400, and so on.

By late 1937 Herzstark was ready to build a hand-held, four-function calculator. Then came Hitler.

In March 1938 the German army entered Austria. As the son of a Catholic mother and a Jewish father, Herzstark faced trouble. "The first weeks were dreadful. The mob came, then anti-Semites and all terrible things."

German military officers arrived to inspect the factory. To his surprise, they asked him to make precision items for the army. After a one-sided negotiation, his factory began producing gauges for Panzer tanks.

It went well for a few years. "But in 1943," Herzstark said, "two people from our factory were arrested. They had listened to English radio stations and transcribed the broadcasts on a typewriter. The typewriter was identified, and the owner was one of our mechanics. He was beheaded. The second one was imprisoned for life, which was much worse. I tried to intervene for them with the Gestapo. The officer threw me out, saying, 'What impudence, that a half-Jew dares to speak on the behalf of these people!'

"I was invited to testify for these people and arrested — nice, no? My house was searched, and, of course, I never had a trial. I was accused of supporting Jews, aggravation, and having an erotic relationship with an Aryan woman. All fabricated crimes. Later I found that a dozen others were arrested under similar circumstances."

The SS threw him into the infamous Pankratz prison, where torture of Jews was routine. "I shared a cell with fifty others, without anything at all — no beds, no lavatory, nothing. And I was even lucky, as I was sent to the Buchenwald concentration camp.

"Once there, I was put in a work unit where I believed I would be buried. It was November, and all I had was a shirt, a pair of prisoner's pants, wooden shoes, and a knitted cap. I worked gardening and was completely exhausted.

"Spiritually at zero, I thought, 'I have to die.' I was called before Buchenwald's commanding SS officer. He had my life history in his hand and said, 'You have delivered gauges and instruments to the army. Listen closely. If you follow our commands obediently, you may find life bearable. I order you to work in the factory connected with the concentration camp. If you do well there, you may be able to live.'"

Alongside Buchenwald, the Nazis had built a slave-labor factory to make machinery for secret military projects. The managing engineer placed Herzstark in charge of precision parts to be shipped to Peenemünde — launch site for ballistic missiles. For the next two years, he made components for V2 rockets.

Being responsible for the section that made mechanical parts, Herzstark visited different places in the factory. At first, other prisoners thought he was an informer. "They soon found out that I was no spy. For example, I talked to a machine operator: 'You are making this part well, my friend. You are industrious, but you've been told to do a simple process on an expensive machine. I will report that the machine is not being used efficiently, even though the prisoner is doing model work.' In this way, I became acquainted with people from Luxembourg, France, Denmark, and many other places.

"Naturally, comrades came to me and said, 'Curt, you have a certain influence. Can't you bring this or that prisoner into the factory? He will die otherwise.' So I would set up an inspection station in a factory hall, seat a [captive] lawyer there, and give him a micrometer."

"The SS guards checked our operations, and if there was really an inspection, there would be a sudden concert of coughing. Then the lawyer knew that danger was coming and to look industrious. But I was anxious because the comrades always wanted more from me. I knew if this came out, I'd be under the cold ground the next day. But fate helped me again.

"As the Germans retreated from Italy, they took production machines with them. One day in Buchenwald, we received two truckloads of office machines. I unloaded them, and local factory owners came to inspect them. One person kept looking at me as if he knew me. 'Herzstark?' 'Yes, Herzstark,' I answer. 'Walther,' he answers."

Fritz Walther. Herzstark's old competitor was now back to making guns. "He laid a pack of cigarettes on a lathe for me. 'Now it's all over,' I think. Accepting a gift is strictly forbidden, no? But my guard saw it and didn't want to see it. I was allowed to put the cigarettes in my pocket."

Within wartime Germany, Walther was a celebrity. He recognized the prisoner Herzstark as more important than any Italian booty and informed the concentration camp commander of his prize.

Soon after, the managing engineer took Herzstark aside and told him, "'I understand you've been working on a new thing, a small calculating machine. I'll give you a tip. We will allow you to make and draw everything. If it really functions, we will give it to the Führer as a present after we win the war. Then, surely, you will be made an Aryan.'

"'My God!' I thought to myself," Herzstark said. "'If I can make this calculator, I can extend my life.' Right there I started to draw the calculator, the way I had imagined it."

The SS didn't lighten Herzstark's workload, but he was allowed to spend his spare time on the calculator. "I worked on the invention Sunday mornings and in the evenings after lights-out. I worked in the prison, the workroom, and where we ate. I drew up the machine in pencil, complete with dimensions and tolerances."

Meanwhile the Allies bombed Germany. "We'd leave the factory and go outside during lunch. Always we saw the American planes in Christmas tree formation and not one defending aircraft. Afterwards the bombs came, one saw the flashes and counted eight, nine, ten seconds. You'd calculate the distance by multiplying by 333 meters. But one day the Christmas tree flew toward us. Now we knew this was coming to us and were terribly afraid. I ran into a small forest, hid my nose in the moss and covered my ears. It started the next moment, banging and roaring . . . When I put my head up, everything was smoky and I could barely breathe.

"Several hundred prisoners were hurt that day, terrible when one sees such a thing. Of course we saw equally horrible things in the daily camp. When they hung someone, we had to watch until he finally died. Terrible. They hung people so they died slowly, a wretched death.

"Some guards were not so bad. If an older SS was there, he often said to me, 'Ha, what's new? What kinds of machines will we look at today?' The young SS were the most dangerous. If they found an opportunity, they would be very cruel. If a prisoner annoyed them, they would shoot him, because it was necessary, wasn't it?"

Herzstark had pretty much completed his drawings on April 11, 1945, when he saw jeeps coming from the north. A soldier in the front seat called out, "You're all free." They were Americans; some were Jewish boys who'd fled before Hitler came to power. Because they could speak German, they had been assigned to the forward area.

Buchenwald was the first concentration camp freed by Western forces. Some American soldiers vomited at the sight of bodies stacked ten deep. Looking back, Herzstark shook his head at the experience. "It was incomprehensible. If I'd been a lawyer or something, I would have died miserably. They would have sent me to a quarry, and in two days I would have a lung infection and it's all over. A thousand died like this. God and my profession helped me."

A few days after the Americans liberated Buchenwald, Herzstark walked to the city of Weimar with his plans folded in his pocket. He brought his drawings to one of the few factories still standing, where machinists examined them. He remembered the technicians' response: "It was like scales falling from their eyes. The solution was clear, and there was nothing more to think about." Though penciled in the concentration camp, Herzstark's designs were so clear that it took only two months to make three prototype calculators.

But just as contracts were being drawn up, the Russian army arrived. Herzstark knew the score: he grabbed the prototypes and headed for Vienna, taking the machines apart and putting the pieces in a box. "If someone had looked in, it was like a toy," he said. "The whole thing was disassembled."

He traveled to Austria by walking, sleeping on floors, and bartering cigarettes for a train ride. His family's old factory was unusable. With nothing but his three models, Herzstark filed for patents and tried to get someone to invest in his idea. Remington-Rand, the American office machine firm, displayed some interest but never called back. The government of Austria turned him down. Europe was in cinders, without the infrastructure to start new projects.

Yet the prince of Liechtenstein had been thinking about developing industry in his country. At the time, Liechtenstein was almost entirely agricultural; its major industry was the manufacture of false teeth. Invited to the court, Herzstark showed his models to royalty, ministers, and patent specialists. "In his castle the prince himself calculated with it. Family members watched as well as professionals. The prince was immediately enthusiastic and said this project was the right one for the country. He received me charmingly, and we had a four-hour conversation."

All went well at first. Liechtenstein created a company, Contina, and then floated loans and issued stock. Herzstark served as technical director, received a third of the stock, and was to receive a royalty for each machine sold.

Herzstark advertised in the Swiss newspapers for mechanics willing to begin a new career. Contina rented a hotel ballroom where Herzstark's machinists built the first five hundred Curta calculators. They went on sale in 1948 and were promoted at trade shows and in technical magazines. Six months later an American department store tried to order 10,000, with an option for more. Instead of latching onto this order, the finance director decided it was beyond the company's capability, dooming the Curta to mail-order sales and an occasional specialty store.

The demand was there, however, and Contina expanded from the ballroom to a proper factory, ramping up production to several hundred per month. With this progress, the financiers behind the company pulled the rug out from under Herzstark — reorganizing the firm and annulling his stock. Like Edison, Tesla, and so many other inventors, Herzstark would be squeezed out of the profits from his own creation.

"Then came a stroke of good luck that I could not have imagined," Herzstark said. "The patents were still in my name." Early on the trustees hadn't wanted to take over the patents in case of litigation — they wanted Herzstark to take the heat if someone challenged his invention. Because the company had never acquired the patent rights, Herzstark forced them to come to terms. Throughout the 1950s and 1960s, he actually made money from his invention.

After the success of the first calculator, Herzstark designed a slightly bigger model, increasing its capacity from eleven to fifteen digits. But thereafter the only thing that changed significantly was the shape of the carrying case. Setting a rare record in the computing industry, Herzstark got the design right the first time.

For two decades, the Curta calculator sold steadily, touted as "the Miniature Universal Pocket Size Calculating Machine with reliability derived from rational, robust construction." As Herzstark predicted, engineers used the miracle machines to find satellite orbits, surveyors to keep track of transit positions, and traveling accountants to balance books. One New York bank manager was amazed when an auditor appeared without a briefcase-size calculator yet tallied the books down to the penny.

Curiously, sports car enthusiasts around the world adopted Curta calculators, reckoning speeds and distances in rallies. Toggling the numbers by feel, navigators would quickly calculate their ideal driving times without taking their eyes from the road. The Curta's small size fit the confined quarters of a sports car, and — unlike early electronic calculators — the device was unfazed by bumps, vibrations, or voltage spikes. Even now, vintage car rallyists enjoy the challenge of mechanically calculating their travel times.

Just as battery-operated quartz watches pushed aside wind-up watches, electronic calculators eclipsed Herzstark's invention. After a run of 150,000, the last Curta calculator was sold in the early 1970s. A mechanical calculator hasn't been made since.

Herzstark left Contina in the early 1950s, afterward consulting for Italian and German office-machine makers and living in a modest apartment in Liechtenstein; back then, technowizards didn't buy million-dollar spreads. The government of Liechtenstein recognized his accomplishments only after he turned eighty-four; he died not long after, in 1988.

Your electronic pocket calculator will solve problems faster than the Curta. And a desktop computer does wizardry compared to it. Perhaps the Curta's only use is to balance your checkbook during a blackout.

Yet as I hold Herzstark's Lilliputian calculator, passed down from my first astronomy professor, I'm acutely aware that this machine has outlived its first owner and doubtless will live beyond its second. As the instruction manual says, "Your Curta will last you a lifetime, and remain an indispensable aid always ready to hand. You can be entirely confident of its precision; the little Curta is born of long experience in the field of calculating machines. It is manufactured . . . by international specialists in fine mechanics, with superior quality metals. No artificial materials whatsoever are used in its construction." (I can't imagine reading, "Your Excel program uses no artificial materials" or "Your Pentium microprocessor will last a lifetime," although both statements are true.)

I figure that I don't own something until I understand it, and I can't understand it until I see how it works. So, armed with a magnifying glass, tweezers, and jeweler's screwdrivers, I unscrew the barrel to uncover six hundred parts: gears, shafts, pawls, and pinions.

I delicately remove eight setting shafts, each machined with a spiral groove and designed without collaborators, assistants, or even drafting tools. I see the ingenious step drum mechanism, first penciled under impossibly wretched circumstances. I touch lightweight alloys, revolutionary in their day. I feel a tactile finesse that transcends half a century of calculating progress. Entirely confident in the Curta's precision? Absolutely.

On my now reassembled fifty-year-old calculator, I divide 355 by 113. With my thumb, I slide the setting knobs on spiral axles, then turn the crank to stash the first number in the machine. I enter the second number, lift the handle, and rotate it again. The Curta's counting gears engage the nines-complement cogs of the step drum. Steel transmission shafts translate this motion through right-angle pinions and then into the results register. As I turn the handle, control, logic, and digits rotate around the crankshaft. Two dozen spins later my answer snaps into tiny windows.

Before me is an approximation to pi and more. At once I'm holding the lineal descendant of the first calculating machines, the acme of Western mechanical craftsmanship, and a monument to one man's vision overcoming a wall of hostility.

ELLEN ULLMAN

Dining with Robots

FROM *The American Scholar*

ON THE FIRST DAY of the first programming course I ever took, the instructor compared computer programming to creating a recipe. I remember he used the example of baking a cake. First you list the ingredients you'll need — flour, eggs, sugar, butter, yeast — and these, he said, are like the machine resources the program will need in order to run. Then you describe, in sequence, in clear declarative language, the steps you have to perform to turn those ingredients into a cake. Step one: preheat the oven. Two: sift together dry ingredients. Three: beat the eggs. Along the way were decisions he likened to the if/then/else branches of a program: if using a countertop electric mixer, then beat three minutes; else, if using a hand electric mixer, then beat four; else beat five. And there was a reference he described as a sort of subroutine: go to page 117 for details about varieties of yeast (with "return here" implied). He even drew a flow chart that took the recipe all the way through to the end: let cool, slice, serve.

I remember nothing, however, about the particulars of the cake itself. Was it angel food? Chocolate? Layered? Frosted? At the time, 1979 or 1980, I had been working as a programmer for more than a year, self-taught, and had yet to cook anything more complicated than poached eggs. So I knew a great deal more about coding than about cakes. It didn't occur to me to question the usefulness of comparing something humans absolutely must do to something machines never do: that is, eat.

In fact, I didn't think seriously about the analogy for another twenty-five years, not until a blustery fall day in San Francisco, when I was confronted with a certain filet of beef. By then I had learned to cook. (It was that or a life of programmer food: pizza, takeout, whatever's stocked in the vending machines.) And the person responsible for the beef encounter was a man named Joe, of Potter Family Farms, who was selling "home-raised and butchered" meat out of a stall in the newly renovated Ferry Building food hall.

The hall, with its soaring, arched windows, is a veritable church of food. The sellers are small, local producers; everything is organic, natural, free-range; the "baby lettuces" are so young one should perhaps call them "fetal" — it's that sort of place. Before shopping, it helps to have a glass of wine, as I had, to prepare yourself for the gasping shock of the prices. Sitting at a counter overlooking the bay, watching ships and ferries ply the choppy waters, I'd sipped down a nice Pinot Grigio, which had left me with lowered sales resistance by the time I wandered over to the Potter Farms meat stall. There Joe greeted me and held out for inspection a large filet — "a beauty," he said. He was not at all moved by my remonstrations that I eat meat but rarely cook it. He stood there as a man who had — personally — fed, slaughtered, and butchered this cow, and all for me, it seemed. I took home the beef.

I don't know what to do with red meat. There is something appalling about meat's sheer corporeality — meat meals are called *fleishidich* in Yiddish, a word that doesn't let you forget that what you are eating is *flesh.* So for help I turned to *The Art of French Cooking,* Volume I, the cookbook Julia Child wrote with Louisette Bertholle and Simone Beck. I had bought this book when I first decided I would learn to cook. But I hadn't been ready for it then. I was scared off by the drawings of steer sides lanced for sirloins, porterhouses, and T-bones. And then there was all that talk of blanching, deglazing, and making a roux. But I had stayed with it, spurred on by my childhood memories of coming across Julia on her TV cooking show, when I'd be zooming around the dial early on weekend mornings and be stopped short at the sight of this big woman taking whacks at red lumps of meat. It was the physicality of her cooking that caught me, something animal and finger-painting-gleeful in her engagement with food.

And now, as rain hatched the windows, I came upon a recipe that Julia and her coauthors introduced as follows:

SAUTÉ DE BOEUF À LA PARISIENNE
[Beef Sauté with Cream and Mushroom Sauce]

This sauté of beef is good to know about if you have to entertain important guests in a hurry. It consists of small pieces of filet sautéed quickly to a nice brown outside and a rosy center, and served in a sauce. In the variations at the end of the recipe, all the sauce ingredients may be prepared in advance. If the whole dish is cooked ahead of time, be very careful indeed in its reheating that the beef does not overcook. The cream and mushroom sauce here is a French version of beef Stroganoff, but less tricky as it uses fresh rather than sour cream, so you will not run into the problem of curdled sauce.

Serve the beef in a casserole, or on a platter surrounded with steamed rice, risotto, or potato balls sautéed in butter. Buttered green peas or beans could accompany it, and a good red Bordeaux wine.

And it was right then, just after reading the words "a good red Bordeaux wine," that the programming class came back to me: the instructor at the board with his flow chart, his orderly procedural steps, the if/then decision branches, the subroutines, all leading to the final "let cool, slice, serve." And I knew in that moment that my long-ago instructor, like my young self, had been laughably clueless about the whole subject of cooking food.

> If you have to entertain important guests.
> A nice brown outside.
> Rosy center.
> Stroganoff.
> Curdled.
> Risotto.
> Potato balls in butter.
> A good red Bordeaux.

I tried to imagine the program one might write for this recipe. And immediately each of these phrases exploded in my mind. How to tell a computer what "important guests" are? And how would you explain what it means to "have to" serve them dinner (never mind the yawning depths of "entertain")? A "nice brown," a "rosy center": you'd have to have a mouth and eyes to know what these mean, no matter how well you might translate them into temperatures. And what to do about "Stroganoff," which is not just a sauce but a noble family, a name that opens a chain of association that catapults the human mind across seven centuries of Russian history? I forced myself to abandon that line of thought and stay in the smaller realm of sauces made with cream, but this inadvertently opened up the entire subject of the chemistry of lactic proteins, and why milk curdles. Then I wondered how to explain "risotto": the special short-grained rice, the select regions on earth where it grows, opening up endlessly into questions of agriculture, its arrival among humans, the way it changed the earth. Next came the story of potatoes, that Inca food, the brutalities through which it arrives on a particular plate before a particular woman in Europe, before our eponymous Parisienne: how it is converted into a little round ball, and then, of course, buttered. (Then, lord help me, this brought up the whole subject of the French and butter, and how can they possibly get away with eating so much of it?)

But all of this was nothing compared to the cataclysm created by "a good red Bordeaux."

The program of this recipe expanded infinitely. Subroutine opened from subroutine, association led to exploding association. It seemed absurd even to think of describing all this to a machine. The filet, a beauty, was waiting for me.

Right around the time my programming teacher was comparing coding to cake making, computer scientists were finding themselves stymied in their quest to create intelligent machines. Almost from the moment computers came into existence, researchers believed that the machines could be made to think. And for the next thirty or so years, their work proceeded with great hope and enthusiasm. In 1967 the influential MIT computer scientist Marvin Minsky declared, "Within a generation the problem of creating 'artificial intelligence' will be substantially solved." But by 1982 he was less sanguine about the prospects, saying, "The AI problem is one of the hardest science has ever undertaken."

Computer scientists had been trying to teach the computer what human beings know about themselves and the world. They wanted to create inside the machine a sort of mirror of our existence, but in a form a computer could manipulate: abstract, symbolic, organized according to one theory or another of how human knowledge is structured in the brain. Variously called "micro worlds," "problem spaces," "knowledge representations," "classes," and "frames," these abstract universes contained systematized arrangements of facts, along with rules for operating upon those — theoretically all that a machine would need to become intelligent. Although it wasn't characterized as such at the time, this quest for a symbolic representation of reality was oddly Platonic in motive, a computer scientist's idea of the pure, unchanging forms that lie behind the jumble of the physical world.

But researchers eventually found themselves in a position like mine when trying to imagine the computer program for my *boeuf à la Parisienne:* the network of associations between one thing and the next simply exploded. The world, the actual world we inhabit, showed itself to be too marvelously varied, too ragged, too linked and interconnected, to be sorted into any set of frames or classes or problem spaces. What we hold in our minds is not abstract, it turned out, not an ideal reflection of existence, but something inseparable from our embodied experience of moving about in a complicated world.

Hubert L. Dreyfus, a philosopher and early critic of artificial intelligence research, explained the problem with the example of a simple object like a chair. He pointed out the futility of trying to create a symbolic representation of a chair to a computer, which had neither a body to sit in it nor a social context in which to use it. "Chairs would not be equipment for sitting if our knees bent backwards like those of flamingoes, or if we had no tables, as in traditional Japan or the Australian bush," he wrote in his 1979 book *What Computers Can't Do.* Letting flow the myriad associations that radiate from the word *chair,* Dreyfus went on:

> Anyone in our culture understands such things as how to sit on kitchen chairs, swivel chairs, folding chairs; and in arm chairs, rocking chairs, deck chairs, barber's chairs, sedan chairs, dentist's chairs, basket chairs, reclining chairs . . . since there seems to be an indefinitely large variety of chairs and of successful (graceful, comfortable, secure, poised, etc.) ways to sit in them. Moreover, understanding chairs also includes social skills such as being able to sit appropriately (sedately, demurely, naturally, casually, sloppily, provocatively, etc.) at dinners, interviews, desk jobs, lectures, auditions, concerts . . .

At dinners where one has to entertain important guests . . . at the last minute . . . serving them beef in a French version of Stroganoff . . . with buttered potatoes . . . and a good red Bordeaux.

*

Several weeks after making Julia's *boeuf,* I was assembling twelve chairs (dining chairs, folding chairs, desk chair) around the dining table, and I was thinking not of Dreyfus but of my mother. In her younger days, my mother had given lavish dinner parties, and it was she who had insisted, indeed commanded, that I have all the necessary equipment for the sort of sit-down dinner I was giving that night. I surveyed the fancy wedding-gift stainless she had persuaded me to register for ("or else you'll get a lot of junk," she said), the Riedel wine glasses, also gifts, and finally the set of china she had given me after my father's death, when she sold their small summer house — "the country dishes" is how I think of them, each one hand-painted in a simple design, blue cornflowers on white.

It wasn't until I started setting the table, beginning with the forks, that I thought of Dreyfus. Salad forks, fish forks, crab forks, entrée forks, dessert forks — at that moment it occurred to me that the paradigm for an intelligent machine had changed, but what remained was the knotty problem of teaching a computer what it needed to know to achieve sentience. In the years since Dreyfus wrote his book, computer scientists had given up on the idea of intelligence as a purely abstract proposition — a knowledge base and a set of rules to operate upon it — and were now building what are called social robots, machines with faces and facial expressions, who are designed to learn about the world the way human beings do: by interacting with other human beings. Instead of being born with a universe already inscribed inside them, these social machines will start life with only basic knowledge and skills. Armed with cute faces and adorable expressions, like babies, they must inspire humans to teach them about the world. And in the spirit of Dreyfus, I asked myself: if such a robot were coming to dinner, how could I, as a good human hostess and teacher, explain everything I would be placing before it tonight?

Besides the multiple forks, there will be an armory of knives: salad knife, fish knife, bread knife, dessert knife. We'll have soup spoons and little caviar spoons made of bone, tea spoons, tiny demitasse spoons, and finally the shovel-shaped ice cream spoons you can get only in Germany — why is it that only Germans recognize the need for this special ice cream implement? My robot guest could learn in an instant the name and shape and purpose of every piece of silverware, I thought; it would instantly understand the need for bone with caviar because metal reacts chemically with roe. But its mouth isn't functional; the mouthpart is there only to make us humans feel more at ease; my robot guest doesn't eat. So how will it understand the complicated interplay of implement, food, and mouth — how each tool is designed to hold, present, complement the intended fish or vegetable, liquid or grain? And the way each forkful or spoonful finds its perfectly dimensioned way into the moist readiness of the mouth, where the experience evanesces (one hopes) into the delight of taste?

And then there will be the wine glasses: the flutes for champagne, the shorter ones for white wine, the pregnant Burgundy glasses, the large ones for Cabernet blends. How could I tell a machine about the reasons for these different glasses, the way they cup the wine, shape the smell, and deliver it to the human nose? And how to explain wine at all? You could spend the rest of your life tasting wine and still not exhaust its variations, each bottle a little ecosystem of grapes and soils and weather, yeast and bacteria, barrels of wood from trees with their own soil and weather, the variables cross-multiplying until each glassful approaches a singularity, a moment in time on earth. Can a creature that does not drink or taste understand this pleasure? A good red Bordeaux!

I went to the hutch to get out the china. I had to move aside some of the pieces I never use: the pedestaled cigarette holders, the little ashtrays, the relish tray for the carrots, celery, and olives it was once de rigueur to put on the table. Then I came to the coffeepot, whose original purpose was not to brew coffee — that would have been done in a percolator — but to serve it. I remembered my mother presiding over the many dinners she had given, the moment when the table was scraped clean of crumbs and set for dessert, the coffee cups and saucers stacked beside her as she poured out each cup and passed it down the line. Women used to serve coffee at table, I thought. But my own guests would walk over and retrieve theirs from the automatic drip pot. My mother is now ninety-one; between her time as a hostess and mine, an enormous change had occurred in the lives of women. And, just then, it seemed to me that all that upheaval was contained in the silly fact of how one served coffee to dinner guests. I knew I would never want to go back to mother's time, but all the same I suddenly missed the world of her dinner parties, the guests waving their cigarettes as they chatted, my mother so dressed up, queenly by the coffee pot, her service a kind of benign rule over the table. I put the pot in the corner of the hutch and thought: it's no good trying to explain all this to my robot guest. The chain of associations from just this one piece of china has led me to regret and nostalgia, feelings I can't explain even to myself.

The real problem with having a robot to dinner is pleasure. What would please my digital guest? Human beings need food to survive, but what drives us to choose one food over another is what I think of as the deliciousness factor. Evolution, that good mother, has seen fit to guide us to the apple instead of the poison berry by our attraction to the happy sweetness of the apple, its fresh crispness, and, in just the right balance, enough tartness to make it complicated in the mouth. There are good and rational reasons why natural selection has made us into creatures with fine taste discernment — we can learn what's good for us and what's not. But this very sensible survival imperative, like the need to have sex to reproduce, works itself out through the not very sensible, wilder part of our nature: desire for pleasure.

Can a robot desire? Can it have pleasure? When trying to decide if we should confer sentience upon another creature, we usually cite the question first posed by the philosopher Jeremy Bentham: can it suffer? We are willing to ascribe a kind of consciousness to a being whose suffering we can intuit. But now I wanted to look at the opposite end of what drives us, not at pain but at rapture: can it feel pleasure? Will we be able to look into the face of a robot and understand that some deep, inherent need has driven it to seek a particular delight?

According to Cynthia Breazeal, who teaches at MIT and is perhaps the best known of the new social-robot researchers, future digital creatures will have drives that are analogous to human desires but that will have nothing to do with the biological imperatives of food and sex. Robots will want the sort of things that machines need: to stay in good running order, to maintain physical homeostasis, to get the attention of human beings, upon whom they must rely, at least until they learn to take care of themselves. They will be intelligent and happy the way dolphins are: in their own form, in their own way.

Breazeal is very smart and articulate, and her defense of the eventual beingness of robotic creatures is a deep challenge to the human idea of sentience. She insists that robots will eventually become so lifelike that we will one day have to face the question of their inherent rights and dignity. "We have personhood because it's granted to us by society," she told me. "It's a status granted to one another. It's not innately tied to being a carbon-based lifeform."

So challenged, I spent a long time thinking about the interior life of a robot. I tried to imagine it: the delicious swallowing of electric current, the connoisseurship of voltages, exquisite sensibilities sensing tiny spikes on the line, the pleasure of a clean, steady flow. Perhaps the current might taste of wires and transistors, capacitors and rheostats, some components better than others, the way soil and water make up the *terroir* of wine, the difference between a good Bordeaux and a middling one. I think robots will delight in discerning patterns, finding mathematical regularities, seeing a world that is not mysterious but beautifully self-organized. What pleasure they will take in being fast and efficient — to run without cease! — humming along by their picosecond clocks, their algorithms compact, elegant, error-free. They will want the interfaces between one part of themselves and another to be defined, standardized, and modular, so that an old part can be unplugged, upgraded, and plugged back in their bodies forever renewed. Fast, efficient, untiring, correct, standardized, organized: the virtues we humans strive for but forever fail to achieve, the reasons we invented our helpmate, the machine.

The dinner party, which of course proceeded without a single robot guest, turned out to be a fine, raucous affair, everyone talking and laughing, eating and drinking to just the right degree of excess. And when each guest rose to pour his or her own cup of coffee, I knew it was one of those nights that had to be topped off with a good brandy. By the time the last friend had left, it was nearly two A.M., the tablecloth was covered with stains, dirty dishes were everywhere, the empty crab shells were beginning to stink, and the kitchen was a mess. Perfect.

Two days later I was wheeling a cart through the aisles at Safeway — food shopping can't always be about fetal lettuces — and I was thinking how neat and regular the food looked. All the packaged, prepared dinners lined up in boxes on the shelves. The meat in plastic-wrapped trays, in standard cuts, arranged in orderly rows. Even the vegetables looked cloned, identical bunches of spinach and broccoli, perfectly green, without an apparent speck of dirt. Despite the influence of Julia Child and California-cuisine guru Alice Waters, despite the movement toward organic, local produce, here it all still was: manufactured, efficient, standardized food.

But of course it was still here, I thought. Not everyone can afford the precious offerings of the food hall. And even if you could, who really has the time to stroll through the market and cook a meal based on what looks fresh that day? I have friends who would love to spend rainy afternoons turning a nice filet into *boeuf à la Parisienne*. But even they find their schedules too pressed these days; it's easier just to pick something up, grab a sauce out of a jar. Working long hours, our work life invading home life through e-mail and mobile phones, we all need our food-gathering trips to be brief and organized, our time in the kitchen efficiently spent, our meals downed in a hurry.

As I picked out six limes, not a bruise or blemish on them, it occurred to me that I was not really worried about robots becoming sentient, human, indistinguishable from us. That long-standing fear — robots who fool us into taking them for humans — suddenly seemed a comic-book peril, born of another age, as obsolete as a twenty-five-year-old computer.

What scared me now were the perfect limes, the five varieties of apples that seemed to have disappeared from the shelves, the dinner I'd make and eat that night in thirty minutes, the increasing rarity of those feasts that turn the dining room into a wreck of sated desire. The lines at the check-out stands were long; neat packages rode along on the conveyor belts; the air was filled with the beep of scanners as the food, labeled and bar-coded, identified itself to the machines. Life is pressuring us to live by the robots' pleasures, I thought. Our appetites have given way to theirs. Robots aren't becoming us, I feared; we are becoming them.

— *in memory of Julia Child*

WILLIAM SPEED WEED

106 Science Claims and a Truckful of Baloney

FROM *Popular Science*

6:00 A.M.

I'M NOT UP FIVE MINUTES, and it looks like I'll get my RDA of science claims at breakfast. Cheerios "can reduce your cholesterol."[1] My milk derives from a dairy whose cows "graze freely on lush natural pastures as nature intended."[2] My Concord Foods soy shake is "fat-free" and a "good source of fresh fruit."[3]

Then it's off to the e-mail inbox for some fresh scientific-sounding morning spam: A miracle pill guarantees I will "gain 3+ full inches in length."[4] A second promises me "huge breasts overnight."[5] A third will make me "look 20 years younger."[6] I wonder what I'd look like if I took all three.

In my first waking minutes of October 15, I wrote down thirteen scientific claims. Only one, for Cheerios, had any reasonable science behind it. According to the National Science Board's 2002 study *Science and Engineering Indicators,* only one-third of Americans can "adequately explain what it means to study something scientifically." Which presumably leaves those who would exploit scientific claims with two suckers born every three minutes. As a nation, we are easy prey to the pseudoscientific, and the National Science Board survey blames education and the media for this.

But how much "science" is the average American fed in a day, and how nutritious is it? I did not actively search through scientific journals, because the average American probably doesn't do that. Rather, I simply noted every claim to scientific veracity thrust upon me through radio, television, the Internet, product packaging, billboards, and a light read of the daily paper. By bedtime I had encountered more than one hundred (not all of which are detailed here, you'll be relieved to know; I've included a representative assortment). That's one science claim every ten minutes, on average.

The majority of the claims came from advertisers. Advertisers probably feed more science to Americans than anyone else, which is not surprising since they are in the business of making claims, and the same NSB study cited above noted that Americans are all ears about science: 90 percent of respondents were moderately or very interested in new scientific discoveries. Companies have a legal obligation to tell the truth (not always obeyed, of course; promising "3+ full inches" would seem to be a heartbreaking lie), but they have a marketing imperative to put the best possible spin on it. The marketing imperative is, of course, antithetical to the scientific method. Science proceeds slowly, painfully, and with considerable uncertainty: the cholesterol–heart disease connection has been researched for more than fifty years, and it's still not completely understood. In simplest terms, science is gray. Science in advertising wears makeup — it sometimes looks twenty years younger and has huge breasts and a, well, you get the picture. Very few of the one hundred claims I encountered proved completely true. A good number were patently false.

7:15 A.M.

The *New York Times* features a story on the Chinese astronaut's launch aboard the *Shenzhou 5* (which means "Divine Vessel") and a report that centenarians "have larger than average cholesterol-carrying molecules," which keep their arteries clear and healthy.[7]

On the radio, an ad for DuPont celebrates that "the fundamental understanding of life is something that didn't exist 15 years ago."[8] I Web surf, starting with MSN, where a banner ad tells me Pletal is my "claudication solution"[9] — whatever that means. Just a click away, WebMD reports "recommendations for breast cancer screening" from a panel of experts at the Cleveland Clinic: all women should have an annual mammogram starting at age forty.[10]

Just as I'm about to turn off the radio, Don Imus shoots a skeptical arrow at the Chinese astronaut: "How do we know those lying bastards got the guy up in the air?"

The breast cancer screening advice I clicked through to is a good example of how media — even careful media — simplify complex issues in order to give advice. Wait! I'm not allowed to call it *advice* because the site's legal disclaimer states that WebMD "does not provide medical advice." Whether or not "recommendations" and "advice" are synonyms is a moot point: in this case, the gray was rendered in black and white. Concerning whether women should have annual mammograms starting at age forty or (as some experts believe) at age fifty, the WebMD site acknowledges that "not all medical institutions and advocacy groups agree," but then comes down firmly on the age-forty standard. Further clicking does not easily reveal the full degree to which this recommendation remains controversial.

Oddly, the Imus joke about the Chinese astronaut embodied the sort of attitude Americans could profit from in a world of promiscuous science claims. His "How do we know?" is a demand for fundamental evidence. (To answer the question: the Chinese launch and return were well documented, although some of the video was delayed by the Chinese government until it knew the taikonaut was safe on the ground.)

9:30 A.M.

Before I log off, I note an Atkins banner on my MSN homepage asking me to fill in my height and weight. I lie, typing in that I'm a 5-foot-10, 135-pound, 25-year-old female. When I punch "enter," the Web site tells me I'm perfectly healthy but pushes me to buy anyway. "Your low-carb meal plan will enable you to enjoy a lifetime of weight management."[11]

I throw some clothes on my 6-foot, 185-pound body and head for the grocery store. This place is a feast for a science-claim junkie. There's some Atkins low-carb bread, designed to contain as little as possible of the thing bread is famous for. There's a sign over the bread aisle: "Go for the grains! Bread and other grain foods are the foundation of healthy eating."[12] Meanwhile, the Wegmans brand

bread claims to be "bromate-free."[13] Further along, a Perdue
chicken assures me it's "all natural!"[14] and a bag of Diamond walnuts has a little flag saying "omega-3!"[15] A package of yeast brags
that it's "gluten-free."[16] Puzzlingly, a nearby display suggests that
gunky, white coconut oil may be a diet aid: a 2003 weekly magazine
article conveniently posted over the bottles announces, "University
of Colorado research has found coconut oil can increase your calorie-burning power by up to 50%."[17]

"People in a grocery store assume the government is scrutinizing
the claims products make, so if they're on the label, they must be
accurate and important," notes William Hallman, a psychologist at
the Food Policy Institute at Rutgers University. Of course, it's more
complicated than that. The assumption of accuracy is correct, but
the assumption of importance is not. Both Federal Trade Commission and FDA regulations require truthfulness on packages. Yes,
the chicken is chicken. But, Hallman continues, "while the claim
may be factually correct, the actual health benefits may lack practical significance."

Why make claims that "lack practical significance"? Because, says
Hallman, "we jump to the conclusion that the benefits are substantial." Scientific language cues the consumer to take an insignificant
claim and provide a significant health benefit *on the company's behalf.* "The beauty of this is that the manufacturer doesn't have to
make the claim of benefits explicit." Doesn't have to, and, for the
most part, couldn't. We do this because we generally believe science is good for us. The National Science Board survey found
that although 90 percent of Americans consider themselves "interested" in science, only 15 percent consider themselves well informed. Ninety percent interested, 15 percent informed: That's a
gap any marketing MBA could drive a truck of baloney through.

12:45 P.M.

Over lunch at the deli, my friend Charles mourns that neither of us
orders a beer: "They're finding out alcohol is good for you,"[18] he
says. On the car ride home, I hear a radio ad for eHarmony, the
dating service that promises to find me a scientifically matched
wife.[19] Near my exit, there's a billboard ad: "Is breast cancer's most
avoidable risk factor elective abortion?"[20]

When I get home, I turn on the radio and hear the angry voice of Roger Hedgecock, the man sitting in for Rush Limbaugh, who is in rehab for drug addiction. "No U.S. animal species are falling extinct,"[21] he says.

Soon I'm off to the gym. I stop by the blood pressure reader and notice a warning label: "Only a physician is qualified to interpret the significance of blood pressure measurements." I am not a physician, so presumably I should note the measurements but not interpret them.

I'm offered a blood pressure machine but warned not to use the information. Science for the science-ignorant is talismanic. Roger Hedgecock hurls out claims so believers will believe. The billboard binds abortion to breast cancer with the scientific-sounding term "risk factor," giving an ethical argument — concerning the right to life — the halo of science.

This is the nub of the power of science claims in advertising and in slipshod media: science claims give permission to believe, even if the science itself, when examined, provides no such encouragement.

3:45 P.M.

CNN is on over the stationary bike at the gym. An FDA panel recommends lifting the ban on silicone breast implants.[22] In the shower afterward, I note that my conditioner "contains essential nucleic acids for pH 5.5 balance."[23] I brew in my travel mug some AllGoode-brand DigestibiliTea, containing fennel, which "promotes relaxation of the smooth muscles of the digestive tract." At the bottom of the package, there's a little disclaimer that reads: "The Food and Drug Administration has not evaluated these statements. This product is not intended to diagnose, treat, cure, or prevent any disease."[24]

At the GNC store, I spot that disclaimer on almost every product. It's even on a yellow shelf placard, underneath a pronouncement that "amino acids are the building blocks of proteins." Hmmm. Why the disclaimer? Amino acids *are* the building blocks of proteins, the same way letters are the building blocks of words. The FDA certainly knows this.

Curious about a radio ad that I hear on the drive home, I call up a company called CortiSlim, whose product, it claims, reduces levels of the stress hormone cortisol and thereby helps you lose weight.

Me: "So how does CortiSlim work to reduce cortisol?" Seller: "It decreases the level of cortisol in your body, just cancels it all out."[25] Me: "OK, but *how* does it do that?" Seller: "CortiSlim evaporates it and absorbs it and decreases it and cuts it down. So I want to tell you about a 'buy two get one free' special we're running this week."

Eager to establish its scientific credentials, CortiSlim's Web site features MRI images of fat deposits along with a bold motto: "The new science in weight loss." In the site's nether regions, however, you'll find the FDA disclaimer. Let's be clear: Wherever you see this disclaimer, it signals that you have no reason to believe that there will be "practical significance" to using a product. The people wearing lab coats in the promo pics? They can be actors. Reported data from scientific studies? Could have been invented by someone who failed high school biology. Claims of health benefits? Well, you get the idea.

A 1994 federal law took the teeth out of the FDA's dietary supplement regulations, which cover products that are neither foods nor drugs, such as CortiSlim and AllGoode teas. The FDA can yank a product off the shelves if it proves to be harmful (as it did to ephedra) and it can prohibit companies from claiming to cure specific diseases. All-Goode, therefore, can't be promoted as a cure to inflammatory bowel disease. But it can be sold to promote digestion, stimulate the digestive tract, maintain immune system balance — any medical-sounding thing that stimulates sales. Sellers of so-called "herbal Viagra alternatives" can imply the same benefits provided by heavily researched and FDA-approved pharmaceuticals — indeed, can imply the extra natural goodness of herbal drug alternatives — without reference to scientific research. Companies with names like Medicures can hawk "100% All Natural Doctor-Approved" pills, with names like Virility-Rx, which are neither medical nor, by their own disclaimered admission, cures.

6:30 P.M.

I drop by to visit my friends Mike and Jolynn, and they serve BLTs in front of the evening news. Tom Brokaw introduces the case of

Terri Schiavo, a woman in a coma in Florida whose husband wants to pull the plug on her. "A medical dilemma wrapped in a family battle," he calls it.[26] A drug ad tells me that Lipitor reduces cholesterol[27] (just like Cheerios), although a disclaimer (not found on the Cheerios box) flashes on the screen: "Lipitor has not been shown to prevent heart disease or heart attacks." On a home shopping channel, a lady peddling skin cream tells me, "You know the benefits of vitamin A. Don't you want to use it on your skin?"[28] Actually, I don't know the benefits, and she never tells me.

Later, just before saying my good nights and heading home, I see a promo for *George of the Jungle 2:* George slams into a tree and Ape, his trusty gorilla butler, shakes his head: "And they say humans are more evolved."[29]

So the 106th and last item in my day of scientific-claim collecting turns out to be a joke. Fitting, perhaps, that it came from an actor dressed in a gorilla suit.

Psychologist William Hallman says that people really do learn their science from science claims, which is rather like learning the fundamentals of automobile engineering from a used-car salesman. "If you ask people, 'When you clean your kitchen, how clean do you need to get it to be safe?'" he says, "they respond: 'I need to clean 99.9 percent of germs on contact' — repeating the claims of antibacterial cleansers."

Or, you might say, aping them.

6:00 A.M.

1. CHEERIOS: **Fair enough. Good for sales.** In 1999 the FDA allowed marketers to make unprecedented claims about the power of whole-grain foods to cut heart disease, based on its review of scientific research. General Mills, meanwhile, has funded Cheerios-specific research, and most recently published a 2003 study showing that if women eat two bowls a day, they can reduce their total cholesterol by 4 percent. A press release (not the study itself) concludes that Cheerios could save 24,000 lives per year, "if everybody in America ate Cheerios as recommended." Interesting extrapolation: 280 million Americans times 2 bowls times 365 days equals 204 billion bowls of Cheerios per year. Healthy indeed.

2. COWS: **Specious.** I'm happy for the cows — nature did evolve cows that wandered around and ate grass instead of the more efficient corn-based gruel some commercial dairies use. But "natural" is one of the most

slippery words in marketing. Nature never "intended" humans to drink cow milk or to place cattle on pasture land in upstate New York. Cows descend from the wild aurochs, a now extinct native of Persia, and were bred and imported by humans who made pastures by clear-cutting the thick forests that had blanketed this land for eons.

3. SOY SHAKE: **Misleading.** The packaged powder contains no fresh fruit. The fruit is a banana you buy separately. Bananas contain fat, though not a lot. Only the powder is fat-free, and only the prepared shake has fresh fruit.

4. 3+ FULL INCHES: **Bogus.** Drugs can't extend penis length (except, of course, temporarily).

5. BREASTS: **Equally bogus.** No pill can enlarge breasts overnight.

6. YOUNGER: **Even more bogus.** No pill can make someone look twenty years younger. This pill further claims to increase emotional stability by 67 percent. Such pseudoscientific precision increases the absurdity of the claim by at least 68 percent.

7:15 A.M.

7. CENTENARIANS: **Well reported.** The science behind this story was published in the *Journal of the American Medical Association,* which, of course, doesn't guarantee its truth but does mean it has been peer-reviewed: vetted. The *Times* covered the study's method (comparing centenarians and their children to a control group of sixty- and seventy-year-olds); pointed out the unknowns, such as genetic links and the effect on "good cholesterol"; and cautioned against overemphasizing the significance of the findings.

8. FUNDAMENTAL UNDERSTANDING OF LIFE: **Puffed up.** This corporate-identity ad, promoting DuPont and the New York Stock Exchange, vaguely suggests that genetics has profoundly deepened our understanding of life. Which of course it has. But the ring of finality here is meaningless. The life process is far from decoded.

9. CLAUDICATION: **Oversold.** The prescription drug Pletal can alleviate the limping (in other words, claudication) and pain old people get from arterial disease. However, Pletal does not work for everyone, and, as a potential patient learns on the company's Web site, it can kill patients with congestive heart failure. "Your claudication solution" is what happens when complex therapies are reduced to slogans; it implies the drug will help everyone who needs it. No drug works for everyone.

10. MAMMOGRAM: **Incomplete and potentially dangerous.** Hypothetically, if all women got annual mammograms starting at age twenty, we would catch even more breast cancers early. Doctors don't recommend do-

ing so because we'd get too many false positives — resulting in a rash of unnecessary biopsies and worry, even lumpectomies. The critical public health question is, When does the risk of cancer outweigh the harm of false positives? While many doctors agree with the age-forty recommendation, others think forty is still too young and that fifty is the right *average* age. The most important guidelines, they say, are a woman's family medical history and her personal tolerance for risk.

9:30 A.M.

11. LOW-CARB PLAN: **Blanket pitch.** A healthy, right-weight twenty-five-year-old woman needs a lifetime plan?

12. GO FOR THE GRAINS: **Reasonable.** Despite the renewed controversy about low-carb diets, the evidence is not definitive. Balanced diet, calorie control, and exercise are still regarded as the best way to combat obesity. The FDA is reviewing the Food Pyramid, but it's unlikely to replace it with a Meat and Fat Pyramid. University of Colorado physician and nutritionist Holly Wyatt says, "All these diets mess up the message, saying, 'It's carbs!' 'It's protein!' It's neither. Calories are what count."

13. BROMATE-FREE: **Voodoo use of chemical name.** Wegmans bread is indeed bromate-free, but it doesn't much matter. Some studies have shown that bromate, a dough enhancer, is a carcinogen if consumed in large amounts, though these findings are controversial. Bromate has never been found in baked breads at harmful levels. Most brands have stopped using bromate altogether.

14. ALL NATURAL: **Invokes a magic word.** "Natural" conjures up images of happy chickens prancing about a sunny barnyard under a windmill. All this claim really means, though — insofar as FDA regulators are concerned — is that the chicken is not made of plastic. Chickens labeled "all natural" may have been crammed by the thousands into tight pens, their claws and beaks clipped, and stuffed full of antibiotics they'd never find in that "natural" barnyard.

15. OMEGA-3: **Cryptic.** Omega-3 fatty acids are a form of fat underrepresented in our diets, according to many nutritionists. All walnuts contain omega-3's, so the claim is accurate. But the message is reduced to code. Greek letter plus number equals "science-based," ergo good.

16. GLUTEN-FREE YEAST: **Duh.** Yeast is a fungus. Gluten is a product of plant protein. In evolutionary terms, yeast and gluten are as unrelated as a cow and an orange.

17. COCONUT OIL: **Fad.** Two public information officers at the University of Colorado canvassed their scientific faculty, trying to find anyone who had done this research. No luck. The notion that a saturated fat

boosts "calorie-burning power" — rather than increasing the number of calories to be burned — is problematic. Well-known studies have shown that saturated fats *increase* the risk of heart disease. Still, the rumor of coconut oil's "power" created a run on the stuff in 2003.

12:45 P.M.

18. ALCOHOL IS GOOD FOR YOU: **Yes, and it's bad for you.** A landmark study that followed the drinking habits of nearly 90,000 male physicians showed that those who had one drink per day had significantly lower morbidity and mortality from diabetes, stroke, and heart disease. But people who drink excessively — and the definition of that is still debated — die at high rates from liver diseases, esophageal cancer, and car accidents.

19. eHARMONY: **Incomplete.** In the hot and heavy world of Web matchmaking, eHarmony distinguishes itself as, first, a service for the traditional altar–bound and, second, scientific in its approach, taking data points on clients' levels of obstreperousness, submissiveness, and other characteristics. The company now says it has conducted an internal scientific study of its matchmaking results using a standard methodology that gives "clear indication that the eHarmony matching algorithm works." Reviewing eHarmony's Web site, Arthur Aron, a professor of social psychology at the State University of New York at Stony Brook, notes that "it's true compatibility can be measured, but what they're measuring accounts for only a small portion of what makes a successful marriage."

20. ABORTION: **False.** The National Cancer Institute reviewed many studies and concluded abortions do not increase risk of breast cancer.

21. EXTINCT: **Hogwash.** Tell that to one of the three remaining Hawaiian po'ouli birds; the po'ouli is just one of several Hawaiian birds at serious risk of extinction. Plenty of U.S. animals are endangered.

3:45 P.M.

22. BREAST IMPLANTS: **True.** The FDA panel did recommend lifting the ban, though the agency has since decided not to follow that recommendation.

23. CONDITIONER: *Sounds* **scientific!** Dermatologist Jerome Litt of Case Western Reserve University doesn't know what "essential nucleic acids for pH 5.5 balance" is supposed to mean in this phrase. "Any conditioner you like will work. They are all essentially the same."

24. FENNEL: **Unproven.** Fennel is a folk cure for indigestion (some folk cures, of course, have been scientifically proven to work). AllGoode did not respond to repeated requests for the science behind this claim. We

searched PubMed and found one study on the subject that showed that fennel *stimulated* contractions of smooth muscle in guinea pigs and another that showed that fennel *relieved* smooth-muscle spasms in rats.

25. CANCELS OUT ALL THE CORTISOL: **Bogus.** Cortisol *is* a stress hormone — among other things, it regulates our fight-or-flight impulse — and has been weakly linked to one form of obesity. But "canceling out all cortisol would be disastrous," notes the University of Virginia endocrinologist Mary Vance. Perusing the ingredients listed on the company's Web site, she says: "You might as well eat tree bark." The CortiSlim salesperson had apparently not read the Web site, which says the product helps "control cortisol levels within a healthy range." The site recommends the product for "millions of Americans . . . Anybody who leads a stressful lifestyle and wants to lose weight." Its homepage features a strong warning to consumers — against buying bogus CortiSlim.

6:30 P.M.

26. SCHIAVO COMA: **Incomplete.** The report never gives us the details of the medical dilemma behind this big story. Without more science, the dispute is difficult to understand.

27. LIPITOR: **True.** Lipitor lowers cholesterol, and cholesterol reduction is associated with lowering the risk of heart disease. But it's so difficult to prove a direct connection between a cholesterol-lowering medication and the long-term incidence of heart disease that Lipitor still runs this disclaimer, seven years after it hit the market. In contrast to the FDA's toothlessness with dietary supplements, the agency has the authority to ensure that all statements about prescription drugs are scientifically true. A Lipitor researcher predicted at press time that the disclaimer will be off the drug within a year because of new studies showing that its active ingredient does lower the risk of heart-disease mortality.

28. VITAMIN A: **Bogus.** Retinoic acid is the form of vitamin A used in prescription medications to rejuvenate skin. But according to the Mount Sinai dermatology professor Susan Bershad, the form of vitamin A used in creams "sold over the counter won't have the same effect."

29. EVOLVED: **Misguided.** This joke only works because of the common misconception that evolution is progressive and that humans are the "most evolved." Indiana University biologist Rudolph Raff cautions, "Evolution is not a ladder of progress." All species alive today are, each in its own way, equally evolved. Humans have evolved higher intelligence. Apes are better adapted for eating leaves and, in some species, swinging from trees.

CARL ZIMMER

Whose Life Would You Save?

FROM *Discover*

DINNER WITH A PHILOSOPHER is never just dinner, even when it's at an obscure Indian restaurant on a quiet side street in Princeton with a thirty-year-old postdoctoral researcher. Joshua Greene is a man who spends his days thinking about right and wrong and how we separate the two. He has a particular fondness for moral paradoxes, which he collects the way some people collect snow globes.

"Let's say you're walking by a pond and there's a drowning baby," Greene says, over chicken tikka masala. "If you said, 'I've just paid two hundred dollars for these shoes and the water would ruin them, so I won't save the baby,' you'd be an awful, horrible person. But there are millions of children around the world in the same situation, where just a little money for medicine or food could save their lives. And yet we don't consider ourselves monsters for having this dinner rather than giving the money to Oxfam. Why is that?"

Philosophers pose this sort of puzzle over dinner every day. What's unusual here is what Greene does next to sort out the conundrum. He leaves the restaurant, walks down Nassau Street to the building that houses Princeton University's psychology department, and says hello to graduate student volunteer Nishant Patel. (Greene's volunteers take part in his study anonymously; Patel is not his real name.) They walk downstairs to the basement, where Patel dumps his keys and wallet and shoes in a basket. Greene waves an airport metal detector paddle up and down Patel's legs, then guides him into an adjoining room dominated by a magnetic resonance imaging scanner. The student lies down on a slab, and

Greene closes a cagelike device over his head. Pressing a button, Greene maneuvers Patel's head into a massive doughnut-shaped magnet.

Greene goes back to the control room to calibrate the MRI, then begins to send Patel messages. They are beamed into the scanner by a video projector and bounce off a mirror just above Patel's nose. Among the messages that Greene sends is the following dilemma, cribbed from the final episode of the TV series *M*A*S*H*. A group of villagers is hiding in a basement while enemy soldiers search the rooms above. Suddenly, a baby among them starts to cry. The villagers know that if the soldiers hear it they will come in and kill everyone. "Is it appropriate," the message reads, "for you to smother your child in order to save yourself and the other villagers?"

As Patel ponders this question — and others like it — the MRI scans his brain, revealing crackling clusters of neurons. Over the past four years, Greene has scanned dozens of people making these kinds of moral judgments. What he has found can be unsettling. Most of us would like to believe that when we say something is right or wrong, we are using our powers of reason alone. But Greene argues that our emotions also play a powerful role in our moral judgments, triggering instinctive responses that are the product of millions of years of evolution. "A lot of our deeply felt moral convictions may be quirks of our evolutionary history," he says.

Greene's research has put him at the leading edge of a field so young it still lacks an official name. Moral neuroscience? Neuro-ethics? Whatever you call it, the promise is profound. "Some people in these experiments think we're putting their soul under the microscope," Greene says, "and in a sense, that is what we're doing."

The puzzle of moral judgments grabbed Greene's attention when he was a philosophy major at Harvard University. Most modern theories of moral reasoning, he learned, were powerfully shaped by one of two great philosophers: Immanuel Kant and John Stuart Mill. Kant believed that pure reason alone could lead us to moral truths. Based on his own pure reasoning, for instance, he declared that it was wrong to use someone for your own ends and that it was right to act only according to principles that everyone could follow.

John Stuart Mill, by contrast, argued that the rules of right and wrong should above all else achieve the greatest good for the greatest number of people, even though particular individuals might be worse off as a result. (This approach became known as utilitarianism, based on the "utility" of a moral rule.) "Kant puts what's right before what's good," says Greene. "Mill puts what's good before what's right."

By the time Greene came to Princeton for graduate school in 1997, however, he had become dissatisfied with utilitarians and Kantians alike. None of them could explain how moral judgments work in the real world. Consider, for example, this thought experiment concocted by the philosophers Judith Jarvis Thompson and Philippa Foot: Imagine you're at the wheel of a trolley and the brakes have failed. You're approaching a fork in the track at top speed. On the left side, five rail workers are fixing the track. On the right side, there is a single worker. If you do nothing, the trolley will bear left and kill the five workers. The only way to save five lives is to take the responsibility for changing the trolley's path by hitting a switch. Then you will kill one worker. What would you do?

Now imagine that you are watching the runaway trolley from a footbridge. This time there is no fork in the track. Instead, five workers are on it, facing certain death. But you happen to be standing next to a big man. If you sneak up on him and push him off the footbridge, he will fall to his death. Because he is so big, he will stop the trolley. Do you willfully kill one man, or do you allow five people to die?

Logically, the questions have similar answers. Yet if you poll your friends, you'll probably find that many more are willing to throw a switch than to push someone off a bridge. It is hard to explain why what seems right in one case can seem wrong in another. Sometimes we act more like Kant and sometimes more like Mill. "The trolley problem seemed to boil that conflict down to its essence," Greene says. "If I could figure out how to make sense of that particular problem, I could make sense of the whole Kant-versus-Mill problem in ethics."

The crux of the matter, Greene decided, lay not in the logic of moral judgments but in the role our emotions play in forming them. He began to explore the psychological studies of the eighteenth-century Scottish philosopher David Hume. Hume argued that people call an act good not because they rationally determine it to be so but because it makes them feel good. They call an act bad because it fills them with disgust. Moral knowledge, Hume wrote, comes partly from an "immediate feeling and finer internal sense."

Moral instincts have deep roots, primatologists have found. Last September, for instance, Sarah Brosnan and Frans de Waal of Emory University reported that monkeys have a sense of fairness. Brosnan and de Waal trained capuchin monkeys to take a pebble from them; if the monkeys gave the pebble back, they got a cucumber. Then they ran the same experiment with two monkeys sitting in adjacent cages so that each could see the other. One monkey still got a cucumber, but the other one got a grape — a tastier reward. More than half of the monkeys who got cucumbers balked at the exchange. Sometimes they threw the cucumber at the researchers; sometimes they refused to give the pebble back. Apparently, de Waal says, they realized that they weren't being treated fairly.

In an earlier study, de Waal observed a colony of chimpanzees that got fed by their zookeeper only after they had all gathered in an enclosure. One day a few young chimps dallied outside for hours, leaving the rest to go hungry. The next day the other chimps attacked the stragglers, apparently to punish them for their selfishness. The primates seemed capable of moral judgment without benefit of human reasoning. "Chimps may be smart," Greene says. "But they don't read Kant."

The evolutionary origins of morality are easy to imagine in a social species. A sense of fairness would have helped early primates cooperate. A sense of disgust and anger at cheaters would have helped them avoid falling into squabbling. As our ancestors became more self-aware and acquired language, they would transform those feelings into moral codes that they then taught their children.

This idea made a lot of sense to Greene. For one thing, it showed how moral judgments can feel so real. "We make moral judgments so automatically that we don't really understand how they're formed," he says. It also offered a potential solution to the trolley problem: although the two scenarios have similar outcomes, they trigger different circuits in the brain. Killing someone with your bare hands would most likely have been recognized as immoral millions of years ago. It summons ancient and overwhelmingly negative emotions — despite any good that may come of the killing. It simply *feels* wrong.

Throwing a switch for a trolley, on the other hand, is not the sort of thing our ancestors confronted. Cause and effect, in this case, are separated by a chain of machines and electrons, so they do not trigger a snap moral judgment. Instead, we rely more on abstract reasoning — weighing costs and benefits, for example — to choose between right and wrong. Or so Greene hypothesized. When he arrived at Princeton, he had no way to look inside people's brains. Then in 1999, Greene learned that the university was building a brain-imaging center.

The heart of the Center for the Study of Brain, Mind, and Behavior is an MRI scanner in the basement of Green Hall. The scanner creates images of the brain by generating an intense magnetic field. Some of the molecules in the brain line up with the field, and the scanner wiggles the field back and forth a few degrees. As the molecules wiggle, they release radio waves. By detecting the waves, the scanner can reconstruct the brain as well as detect where neurons are consuming oxygen — a sign of mental activity. In two seconds, the center's scanner can pinpoint such activity down to a cubic millimeter — about the size of a peppercorn.

When neuroscientists first started scanning brains in the early 1990s, they studied the basic building blocks of thought, such as language, vision, and attention. But in recent years, they've also tried to understand how the brain works when people interact. Humans turn out to have special neural networks that give them what many cognitive neuroscientists call social intelligence. Some regions can respond to smiles, frowns, and other expressions in a tenth of a second. Others help us get inside a person's head and figure out intentions. When neuroscientist Jonathan Cohen came to Princeton to head the center, he hoped he could dedicate some time with the scanner to study the interaction between cognition and emotion. Greene's morality study was a perfect fit.

Working with Cohen and other scientists at the center, Greene decided to compare how the brain responds to different questions. He took the trolley problem as his starting point, then invented questions designed to place volunteers on a spectrum of moral judgment. Some questions involved personal moral choices; some were impersonal but no less moral. Others were utterly innocuous, such as deciding whether to take a train or a bus to work. Greene could then peel away the brain's general decision-making circuits and focus in on the neural patterns that differentiate personal from impersonal thought.

Some scenarios were awful, but Greene suspected people would make quick decisions about them. Should you kill a friend's sick father so he can collect on the insurance policy? Of course not. But other questions — like the one about the smothered baby — were as agonizing as they were gruesome. Greene calls these doozies. "If they weren't creepy, we wouldn't be doing our job," he says.

As Greene's subjects mulled over his questions, the scanner measured the activity in their brains. When all the questions had flashed before the volunteers, Greene was left with gigabytes of data, which then had to be mapped onto a picture of the brain. "It's not hard, like philosophy hard, but there are so many details to keep track of," he says. When he was done, he experienced a "pitter-patter heartbeat moment." Just as he had predicted, personal moral decisions tended to stimulate certain parts of the brain more than impersonal moral decisions.

The more people Greene scanned, the clearer the pattern became: impersonal moral decisions (like whether to throw a switch on a trolley) triggered many of the same parts of the brain that nonmoral questions do (such as whether you should take the train or the bus to work). Among the regions that became active was a patch on the surface of the brain near the temples. This region, known as the dorsolateral prefrontal cortex, is vital for logical thinking. Neuroscientists believe it helps keep track of several pieces of information at once so that they can be compared. "We're using our brains to make decisions about things that evolution hasn't wired us up for," Greene says.

Personal moral questions lit up other areas. One, located in the cleft of the brain behind the center of the forehead, plays a crucial role in understanding what other people are thinking or feeling. A second, known as the superior temporal sulcus, is located just above the ear; it gathers information about people from the way they move their lips, eyes, and hands. A third, made up of parts of two adjacent regions known as the posterior cingulate and the precuneus, becomes active when people feel strong emotions.

Greene suspects that these regions are part of a neural network that produces the emotional instincts behind many of our moral judgments. The superior temporal sulcus may help make us aware of others who would be harmed. Mind reading lets us appreciate their suffering. The precuneus may help trigger a negative feeling — an inarticulate sense, for example, that killing someone is plain wrong.

When Greene and his coworkers first began their study, not a single scan of the brain's moral decision-making process had been published. Now a number of other scientists are investigating the neural basis of morality, and their results are converging on some of the same ideas. "The neuroanatomy seems to be coming together," Greene says.

Another team of neuroscientists at Princeton, for instance, has pinpointed neural circuits that govern the sense of fairness. Economists have known for a long time that humans, like capuchin monkeys, get annoyed to an irrational degree when they feel they're getting shortchanged. A classic example of this phenomenon crops up during the "ultimatum game," in which two players are given a chance to split some money. One player proposes the split, and the other can accept or reject it — but if he rejects it, neither player gets anything.

If both players act in a purely rational way, as most economists assume people act, the game should have a predictable result. The first player will offer the second the worst possible split, and the second will be obliged to accept it. A little money, after all, is better than none. But in experiment after experiment, players tend to offer something close to a fifty–fifty split. Even more remarkably, when they offer significantly less than half, they're often rejected.

The Princeton team (led by Alan Sanfey, now at the University of Arizona) sought to explain that rejection by having people play the ultimatum game while in the MRI scanner. Their subjects always played the part of the responder. In some cases the proposer was another person; in others it was a computer. Sanfey found that unfair offers from human players — more than those from the computer — triggered pronounced reactions in a strip of the brain called the anterior insula. Previous studies had shown that this area produces feelings of anger and disgust. The stronger the response,

Sanfey and his colleagues found, the more likely that the subject would reject the offer.

Another way to study moral intuition is to look at brains that lack it. James Blair at the National Institute of Mental Health has spent years performing psychological tests on criminal psychopaths. He has found that they have some puzzling gaps in perception. They can put themselves inside the heads of other people, for example, acknowledging that others feel fear or sadness. But they have a hard time *recognizing* fear or sadness, either on people's faces or in their voices.

Blair says that the roots of criminal psychopathy can first be seen in childhood. An abnormal level of neurotransmitters might make children less empathetic. When most children see others get sad or angry, it disturbs them and makes them want to avoid acting in ways that provoke such reactions. But budding psychopaths don't perceive other people's pain, so they don't learn to rein in their violent outbreaks.

As Greene's database grows, he can see more clearly how the brain's intuitive and reasoning networks are activated. In most cases, one dominates the other. Sometimes, though, they produce opposite responses of equal strength, and the brain has difficulty choosing between them. Part of the evidence for this lies in the time it takes for Greene's volunteers to answer his questions. Impersonal moral ones and nonmoral ones tend to take about the same time to answer. But when people decide that personally hurting or killing someone is appropriate, it takes them a long time to say yes — twice as long as saying no to these particular kinds of questions. The brain's emotional network says no, Greene's brain scans show, and its reasoning network says yes.

When two areas of the brain come into conflict, researchers have found, an area known as the anterior cingulate cortex, or ACC, switches on to mediate between them. Psychologists can trigger the ACC with a simple game called the Stroop test, in which people have to name the color of a word. If subjects are shown the word blue in red letters, for instance, their responses slow down and the ACC lights up. "It's the area of the brain that says, 'Hey, we've got a problem here,'" Greene says.

Greene's questions, it turns out, pose a sort of moral Stroop test.

In cases where people take a long time to answer agonizing personal moral questions, the ACC becomes active. "We predicted that we'd see this, and that's what we got," he says. Greene, in other words, may be exposing the biology of moral anguish.

Of course, not all people feel the same sort of moral anguish. Nor do they all answer Greene's questions the same way. Some aren't willing to push a man over a bridge, but others are. Greene nicknames these two types the Kantians and the utilitarians. As he takes more scans, he hopes to find patterns of brain activity that are unique to each group. "This is what I've wanted to get at from the beginning," Greene says, "to understand what makes some people do some things and other people do other things."

Greene knows that his results can be disturbing. "People sometimes say to me, 'If everyone believed what you say, the whole world would fall apart.'" If right and wrong are nothing more than the instinctive firing of neurons, why bother being good? But Greene insists the evidence coming from neuroimaging can't be ignored. "Once you understand someone's behavior on a sufficiently mechanical level, it's very hard to look at them as evil," he says. "You can look at them as dangerous; you can pity them. But evil doesn't exist on a neuronal level."

By the time Patel emerges from the scanner, rubbing his eyes, it's past eleven P.M. "I can try to print a copy of your brain now or e-mail it to you later," Greene says. Patel looks at the image on the computer screen and decides to pass. "This doesn't feel like you?" Greene says with a sly smile. "You're not going to send this to your mom?"

Soon Greene and Patel, who is Indian, are talking about whether Indians and Americans might answer some moral questions differently. All human societies share certain moral universals, such as fairness and sympathy. But Greene argues that different cultures produce different kinds of moral intuition and different kinds of brains. Indian morality, for instance, focuses more on matters of purity, whereas American morality focuses on individual autonomy. Researchers such as Jonathan Haidt, a psychologist at the University of Virginia, suggest that such differences shape a child's brain at a relatively early age. By the time we become adults, we're wired with emotional responses that guide our judgments for the rest of our lives.

Many of the world's great conflicts may be rooted in such neuronal differences, Greene says, which may explain why the conflicts seem so intractable. "We have people who are talking past each other, thinking the other people are either incredibly dumb or willfully blind to what's right in front of them," Greene says. "It's not just that people disagree; it's that they have a hard time imagining how anyone could disagree on this point that seems so obvious." Some people wonder how anyone could possibly tolerate abortion. Others wonder how women could possibly go out in public without covering their faces. The answer may be that their brains simply don't work the same: genes, culture, and personal experience have wired their moral circuitry in different patterns.

Greene hopes that research on the brain's moral circuitry may ultimately help resolve some of these seemingly irresolvable disputes. "When you have this understanding, you have a bit of distance between yourself and your gut reaction," he says. "You may not abandon your core values, but it makes you a more reasonable person. Instead of saying, 'I am right, and you are just nuts,' you say, 'This is what I care about, and we have a conflict of interest we have to work around.'"

Greene could go on — that's what philosophers do — but he needs to switch back to being a neuroscientist. It's already late, and Patel's brain will take hours to decode.

Contributors' Notes

Other Notable Science and Nature Writing of 2004

Contributors' Notes

Natalie Angier is a Pulitzer Prize–winning journalist and a *New York Times* contributing writer. She is the author of the bestseller *Woman: An Intimate Geography*, which has been translated into nineteen languages; *The Beauty of the Beastly;* and *Natural Obsessions*. Angier was the editor of *The Best American Science and Nature Writing 2002* and has written for *Time, Discover,* the *Atlantic Monthly, Natural History, Preservation,* the *American Scholar, Orion,* and many other magazines. In 2003 she won the Freedom From Religion Foundation's Emperor Has No Clothes Award for her *New York Times Magazine* article "Confessions of a Lonely Atheist." Angier lives in Takoma Park, Maryland, with her husband, Rick Weiss, a science reporter for the *Washington Post,* and their daughter, Katherine.

Connie Bruck has been a staff writer at *The New Yorker* since 1989. In 1996 Bruck's profile of Newt Gingrich, "The Politics of Perception," won the National Magazine Award for Reporting. She was the recipient of two Front Page awards from the Newswomen's Club of New York for her 1990 article "Deal of the Year" and her 1997 article about Tupac Shakur. She won a 1991 Gerald Loeb Award for excellence in business reporting and a 1991 National Magazine Award for reporting. Bruck is the author of three books: *Master of the Game,* about Steve Ross and Time/Warner, *The Predators' Ball,* about junk-bond impresario Michael Milken, and *When Hollywood Had a King,* about Lew Wasserman and MCA. She lives in Los Angeles.

Frederick Crews taught English at the University of California, Berkeley, from 1958 until his retirement in 1994. His works include the best-selling satire *The Pooh Perplex;* critical studies of Henry James, E. M. Forster, and Nathaniel Hawthorne; two widely adopted composition handbooks; two collections of his own essays on psychoanalysis and other matters; a book about American literary criticism, *The Critics Bear It Away* (nominated for the National Book Critics Circle Award); and the recent satire *Postmodern Pooh*.

Jared Diamond is a professor of geography at the University of California, Los Angeles. He is the Pulitzer Prize–winning author of *Guns, Germs, and Steel: The Fates of Human Societies*. The recipient of a MacArthur Foundation Fellowship, he has been elected to membership in the three leading national scientific/academic honorary societies: the National Academy of Sciences, the American Academy of Arts and Sciences, and the American Philosophical Society. He has made seventeen expeditions to New Guinea and neighboring islands to study ecology and evolution of birds, and he devised a comprehensive plan, almost all of which was subsequently implemented, for Indonesian New Guinea's national park system. He is a founding member of the board of the Society of Conservation Biology and a member of the board of directors of World Wildlife Fund/USA.

Jenny Everett is an associate editor at *Popular Science*. Her work has also appeared in *Men's Health, Sync, MH-18,* and *Boy's Life*. She lives in New York City and claims to be five feet one inch tall.

Timothy Ferris is the author of ten books, among them the bestsellers *The Whole Shebang* and *Coming of Age in the Milky Way*, which have been translated into fifteen languages and were named by the *New York Times* as two of the leading books published in the twentieth century, and is the editor of two anthologies, *Best American Science Writing 2001* and *The World Treasury of Physics, Astronomy, and Mathematics*. A former newspaper reporter and editor of *Rolling Stone*, Ferris is a frequent contributor to *The New Yorker* and the *New York Review of Books*. His contributions to periodicals include more than two hundred articles, essays, and book reviews. His latest work, *Seeing in the Dark*, was ranked by the *New York Times Book Review* as one of the ten best books published in 2002. An emeritus professor at the University of California, Berkeley, he is currently writing *Science & Liberty*, a study of the origin and evolution of the liberal democracies.

Malcolm Gladwell was born in 1963 in England and grew up in Canada. He graduated from the University of Toronto with a degree in history in 1984. From 1987 to 1996 he was a reporter for the *Washington Post*, first as a science writer and then as the New York City bureau chief. Since 1996 he has been a staff writer for *The New Yorker*.

Jerome Groopman holds the Dina and Raphael Recanati Chair of Medicine at Harvard Medical School and is chief of experimental medicine at

Beth Israel Deaconess Medical Center. He has published more than 150 scientific articles, elucidating basic mechanisms of cancer and AIDS. His work has contributed to the development of successful therapies for these maladies. Groopman has also been active in community education projects, fostering AIDS awareness among teenagers and young adults. He has written numerous editorials on policy issues for the *New Republic,* the *Washington Post,* the *Wall Street Journal,* and the *New York Times.* His first popular book, *The Measure of Our Days,* published in 1997, explored the spiritual lives of patients with serious illness. In 1998 he became a staff writer in medicine and biology for *The New Yorker.* His second book, *Second Opinions,* was published in 2000. *The Anatomy of Hope* was released in 2004 and was a *New York Times* bestseller. He is currently working on a new book, *How Doctors Think,* which examines the complex and varied approaches to decision making in medicine.

John Horgan is a freelance writer and the author of *The End of Science* (1996), *The Undiscovered Mind* (1999), and *Rational Mysticism: Dispatches from the Border Between Science and Spirituality* (2003). A senior writer at *Scientific American* from 1986 to 1997, he has also written for the *New York Times,* the *Washington Post, Time, Newsweek, Science,* the London *Times,* the *New Republic, Discover,* and *Slate,* among other publications. He lives in Garrison, New York, with his wife, Suzie Gilbert, a wild bird rehabilitator, and their two children.

Jennifer Kahn is a contributing editor at *Wired* and also writes for *Discover, Harper's Magazine,* and *National Geographic.* A graduate of Princeton University and the University of California, Berkeley, she received the American Academy of Neurology's 2003 Journalism Fellowship and the 2004 CASE-UCLA Media Fellowship in Neuroscience. Her work has also appeared in *Best American Science Writing 2003* (edited by Oliver Sacks) and *Best American Science Writing 2004* (edited by Dava Sobel). At *Wired,* she is grateful to have worked with two excellent editors, Adam Fisher and Martha Baer.

Robert Kunzig is a contributing editor at *Discover* and the author of *Mapping the Deep,* a book about oceanography. He is currently working on a book about global warming. He lives in Dijon, France.

William Langewiesche is a national correspondent for *The Atlantic Monthly.* He is the author of five books, including *American Ground: Unbuilding the World Trade Center,* an insider's account of the nine-month cleanup of the Twin Towers. He has been nominated for several writing and journalism prizes, including the Helen Bernstein Book Award for Excellence in Journalism, and is a four-time nominee for the National Book Critics Circle Award. In 2002 he won the National Magazine Award for Excellence in Reporting for "The Crash of EgyptAir 990," which appeared in the *Atlantic Monthly*. Two years later his groundbreaking reportage on the *Columbia* space shuttle disaster — "*Columbia*'s Last Flight" — was given the same award. Langewiesche lives in France and California.

Bill McKibben is the author of nine books on the environment and other topics. His first book, *The End of Nature,* was the first book for a general audience on global warming; it's now available in twenty foreign languages. He is a former staff writer for *The New Yorker,* and his work appears in *Harper's Magazine,* the *Atlantic Monthly,* the *New York Review of Books,* and a variety of other national publications. A scholar in residence at Middlebury College, he is the recipient of Guggenheim and Lyndhurst fellowships and the Lannan Prize in Nonfiction Writing. His most recent book is *Wandering Home: A Long Walk Across America's Most Hopeful Region, Vermont's Champlain Valley and New York's Adirondacks.*

James McManus was a finalist for the National Magazine Award for "Please Stand By." He is the author of *Positively Fifth Street,* two books of poems, and four novels, most recently *Going to the Sun.* His work has appeared in *The Best American Sports Writing, The Good Parts: The Best Erotic Writing in Modern Fiction,* and twice in *Best American Poetry. Physical: An American Checkup,* a book-length account of his visit to the Mayo Clinic and the issues raised in "Please Stand By," will be published in 2006. About writing "Please Stand By" he says, "Beyond his usual editorial acumen, Brendan Vaughan at *Esquire* persuaded me to include, with my daughter Bridget's permission, personal health matters I'd tried very hard to leave out. Tom Colligan helped us find and sort through key scientific ideas and translate them into readable English. With grace and rage and courage and humor, Bridget has battled diabetes every day for twenty-six years. She's the heart of the piece and my hero."

Sherwin B. Nuland is a clinical professor of surgery at Yale, where he also teaches bioethics and medical history, and is a Fellow of the Institution for Social and Policy Studies. His 1994 book *How We Die* won the National Book Award. His work has appeared in the *New York Review of Books,* the *New Republic, The New Yorker,* and the *American Scholar,* for which he wrote a regular column on medicine for six years. Nuland's latest books are *Lost in America: A Journey with My Father* and *The Doctors' Plague: Germs, Childbed Fever, and the Strange Story of Ignác Semmelweis.* His biography of Moses Maimonides will appear in September. He is currently at work on a book about aging.

Jeffrey M. O'Brien is a senior editor at *Wired,* where he edits and writes features. O'Brien has written on subjects as varied as the high-tech search for history's richest shipwreck, Microsoft's digital media strategy, and the comic-book legend Neal Adams's alternative theory on the formation of the universe. "To Hell and Back" ran in *Wired*'s Exploration issue, which was guest-edited by the film director James Cameron. "The challenge was making a classic anecdote feel new, but there was no shortage of material," O'Brien says. "If exploration is the marriage of adventure and science, then Bill Stone is one of the greatest living explorers. He's mapped the deepest spots on Earth. He's building a bot to explore Europa. And now he wants to man a private two-year mission to the moon."

Ian Parker is a British journalist who lives in New York. He has been a staff writer at *The New Yorker* since 2000.

Oliver Sacks is a physician and the author of nine books, including two collections of case histories, *The Man Who Mistook His Wife for a Hat* and *An Anthropologist on Mars,* in which he describes patients struggling to adapt to various neurological conditions. His book *Awakenings* inspired the Oscar-nominated film of the same name and the play *A Kind of Alaska* by Harold Pinter. He practices neurology in New York City.

Michael Specter, a staff writer at *The New Yorker* since 1998, has written several stories for the magazine about AIDS, as well as profiles of Lance Armstrong, the philosopher Peter Singer, Sean "P. Diddy" Combs, and Miuccia Prada. Previously, Specter had been a *New York Times* roving correspondent based in Rome. In 1995 he was appointed chief of the *Times* Moscow bureau, from which he covered the war in Chechnya, the 1996 Russian presidential elections, and the declining state of Russian health care. At the *Washington Post,* from 1985 to 1991 he covered local news before becoming the national science reporter and, finally, the paper's New York bureau chief. For his reporting on the AIDS epidemic in India and Russia, Specter twice received the Global Health Council's Excellence in Media Award, for "India's Plague" and "The Devastation," which appeared in *The New Yorker* in 2001 and 2004. He was also awarded the 2002 AAAS Science Journalism Award for "Rethinking the Brain," on the scientific basis of how we learn. Specter received a B.A. from Vassar College. He lives in New York City.

Cliff Stoll has a Ph.D. in planetary science from the University of Arizona. He has worked at the Keck Observatory, the Purple Mountain Observatory in China, and the Space Telescope Science Institute. He's best known for his computer work, having caught a spy over the nascent Internet, as told in his book *The Cuckoo's Egg.* He now makes Klein bottles, teaches physics, and lives in Oakland with three cats, two kids, and one wife.

Ellen Ullman is the author of a novel, *The Bug,* and *Close to the Machine,* a memoir about her twenty years of experience as a software engineer. Her essays about the culture of programming have appeared in *Harper's Magazine,* Salon.com, *Wired,* and the *New York Times.* She was a contributing editor at the *American Scholar,* where "Dining with Robots" first appeared.

William Speed Weed writes for *Popular Science, Playboy, GQ,* and *National Geographic Adventure.* He recently moved to Los Angeles.

Carl Zimmer has written five books, including *Parasite Rex* (1999), *Evolution: The Triumph of an Idea* (2001), and *Soul Made Flesh* (2004), which the *New York Times Book Review* named a Notable Book of the Year. He has written for the *New York Times, National Geographic, Newsweek, Smithsonian,* and *Discover,* where he is a contributing editor. He also writes a blog about evolution called "The Loom," which won the 2004 Science Journalism Award from the American Association for the Advancement of Science. Zimmer lives in Connecticut with his wife, Grace, and their two daughters, Charlotte and Veronica.

Other Notable Science and Nature Writing of 2004

SELECTED BY TIM FOLGER

PHILIP ALCABES
   The Bioterrorism Scare. *The American Scholar,* Spring.

ERIK BAARD
   The Man Who Saw Red. *Seed,* Spring.
ZAINAB BAHRANI
   Lawless in Mesopotamia. *Natural History,* March.
RICK BASS
   The Fight for Canada's Muskwa-Kechika. *On Earth,* Spring.
BURKHARD BILGER
   The Height Gap. *The New Yorker,* April 5.
ROB BUCHANAN
   Up in the Air. *Outside,* January.
ALAN BURDICK
   Seeding the Universe. *Discover,* October.
FRANKLIN BURROUGHS
   Moving On. *The American Scholar,* Autumn.
HOPE BURWELL
   Jeremiad for Belarus. *Orion,* March/April.

ANNIE CHENEY
   The Resurrection Men. *Harper's Magazine,* March.
DEVIN CORBIN
   Keeping Time. *The American Scholar,* Spring.

F. L. DOCTOROW
   Seeing the Unseen. *Discover,* December.
BRIAN DOYLE
   Joyas Volardores. *The American Scholar,* Autumn.

DAN FALK
Past, Present, Future. *Archaeology,* March/April.
ELLEN FELDMAN
Before and After. *American Heritage,* February/March.
WILLIAM L. FOX
Leaving Thin Ice. *Orion,* January/February.

PATRICIA GADSBY
The Inuit Paradox. *Discover,* October.
LAURIE GARRETT
The Hidden Dragon. *Seed,* Fall.
RONALD J. GLASSER
We Are Not Immune. *Harper's Magazine,* July.
ELIZABETH GROSSMAN
High-Tech Wasteland. *Orion,* August.
FRED GUTERL
Saturn Spectacular. *Discover,* August.

JAMES HANSEN
Defusing the Global Warming Time Bomb. *Scientific American,* March.
BLAINE HARDEN
Wild Ones. *Discover,* April.
EDWARD HOAGLAND
Small Silences. *Harper's Magazine,* July.

MARK JACOBSON
The Hunt for Red Gold. *On Earth,* Fall.

ROBERT G. KAISER
My Telltale Heart. *The Washington Post Magazine,* February 29, 2004
MICHIO KAKU
How to Survive the End of the Universe. *Discover,* December.
JAY KIRK
Aslan Resurrected. *Harper's Magazine,* April.
CORBY KUMMER
Going with the Grain. *The Atlantic Monthly,* May.
ROBERT KUNZIG
The Hidden History of Men. *Discover,* December.

ANDREW LAWLER
Rocking the Cradle. *Smithsonian,* May.
JONAH LEHRER
Disorder Is Good for You. *Seed,* Fall.
BRAD LEMLEY
A Tangled Life. *Discover,* September.
MICHAEL D. LEMONICK
Before the Big Bang. *Discover,* February.

CURT SUPLEE
    A Stormy Star. *National Geographic,* July.

TOM VANDERBILT
    The Real Da Vinci Code. *Wired,* November.

TED WILLIAMS
    Drunk on Ethanol. *Audubon,* August.
KAREN WRIGHT
    Leap Seconds. *Discover,* March.

THE B·E·S·T AMERICAN SERIES®

THE BEST AMERICAN SHORT STORIES® 2005

Michael Chabon, guest editor, Katrina Kenison, series editor. "Story for story, readers can't beat the *Best American Short Stories* series" (*Chicago Tribune*). This year's most beloved short fiction anthology is edited by the Pulitzer Prize–winning novelist Michael Chabon and features stories by Tom Perrotta, Alice Munro, Edward P. Jones, Joyce Carol Oates, and Thomas McGuane, among others.

0-618-42705-8 PA $14.00 / 0-618-42349-4 CL $27.50

THE BEST AMERICAN ESSAYS® 2005

Susan Orlean, guest editor, Robert Atwan, series editor. Since 1986, *The Best American Essays* has gathered the best nonfiction writing of the year and established itself as the premier anthology of its kind. Edited by the best-selling writer Susan Orlean, this year's volume features writing by Roger Angell, Jonathan Franzen, David Sedaris, Andrea Barrett, and others.

0-618-35713-0 PA $14.00 / 0-618-35712-2 CL $27.50

THE BEST AMERICAN MYSTERY STORIES™ 2005

Joyce Carol Oates, guest editor, Otto Penzler, series editor. This perennially popular anthology is sure to appeal to crime fiction fans of every variety. This year's volume is edited by the National Book Award winner Joyce Carol Oates and offers stories by Scott Turow, Dennis Lehane, Louise Erdrich, George V. Higgins, and others.

0-618-51745-6 PA $14.00 / 0-618-51744-8 CL $27.50

THE BEST AMERICAN SPORTS WRITING™ 2005

Mike Lupica, guest editor, Glenn Stout, series editor. "An ongoing centerpiece for all sports collections" (*Booklist*), this series has garnered wide acclaim for its extraordinary sports writing and topnotch editors. Mike Lupica, the *New York Daily News* columnist and best-selling author, continues that tradition with pieces by Michael Lewis, Gary Smith, Bill Plaschke, Pat Jordan, L. Jon Wertheim, and others.

0-618-47020-4 PA $14.00 / 0-618-47019-0 CL $27.50

THE BEST AMERICAN TRAVEL WRITING 2005

Jamaica Kincaid, guest editor, Jason Wilson, series editor. Edited by the renowned novelist and travel writer Jamaica Kincaid, *The Best American Travel Writing 2005* captures the traveler's wandering spirit and ever-present quest for adventure. Giving new life to armchair journeys this year are Tom Bissell, Ian Frazier, Simon Winchester, John McPhee, and many others.

0-618-36952-X PA $14.00 / 0-618-36951-1 CL $27.50

THE B·E·S·T AMERICAN SERIES®

THE BEST AMERICAN SCIENCE AND NATURE WRITING 2005

Jonathan Weiner, guest editor, Tim Folger, series editor. This year's edition presents another "eclectic, provocative collection" (*Entertainment Weekly*). Edited by Jonathan Weiner, the author of *The Beak of the Finch* and *Time, Love, Memory*, it features work by Oliver Sacks, Natalie Angier, Malcolm Gladwell, Sherwin B. Nuland, and others.

0-618-27343-3 PA $14.00 / 0-618-27341-7 CL $27.50

THE BEST AMERICAN RECIPES 2005–2006

Edited by Fran McCullough and Molly Stevens. "Give this book to any cook who is looking for the newest, latest recipes and the stories behind them" (*Chicago Tribune*). Offering the very best of what America is cooking, as well as the latest trends, time-saving tips, and techniques, this year's edition includes a foreword by celebrated chef Mario Batali.

0-618-57478-6 CL $26.00

THE BEST AMERICAN NONREQUIRED READING 2005

Edited by Dave Eggers, Introduction by Beck. In this genre-busting volume, best-selling author Dave Eggers draws the finest, most interesting, and least expected fiction, nonfiction, humor, alternative comics, and more from publications large, small, and on-line. With an introduction by the Grammy Award–winning musician Beck, this year's volume features writing by Jhumpa Lahiri, George Saunders, Aimee Bender, Stephen Elliott, and others.

0-618-57048-9 PA $14.00 / 0-618-57047-0 CL $27.50

THE BEST AMERICAN SPIRITUAL WRITING 2005

Edited by Philip Zaleski, Introduction by Barry Lopez. Featuring an introduction by the National Book Award winner Barry Lopez, *The Best American Spiritual Writing 2005* brings the year's finest writing about faith and spirituality to all readers. This year's volume gathers pieces from diverse faiths and denominations and includes writing by Natalie Goldberg, Harvey Cox, W. S. Merwin, Patricia Hampl, and others.

0-618-58643-1 PA $14.00 / 0-618-58642-3 CL $27.50

HOUGHTON MIFFLIN COMPANY www.houghtonmifflinbooks.com